AF892266

Herbal Medicine and Botanical Medical Fads

Herbal Medicine and Botanical Medical Fads

Frank Hoffmann, PhD, MLS
Martin Manning, MLS

Routledge
Taylor & Francis Group
New York London

© 2002 by The Haworth Press, Inc. All rights reserved. No part of this work may be reproduced or utilized in any form or by any means, electronic or mechanical, including photocopying, microfilm, and recording, or by any information storage and retrieval system, without permission in writing from the publisher.

First published by

The Haworth Press, Inc., 10 Alice Street, Binghamton, NY 13904-1580.

This edition published 2012 by Routledge

Routledge
Taylor & Francis Group
711 Third Avenue
New York, NY 10017

Routledge
Taylor & Francis Group
2 Park Square, Milton Park
Abingdon, Oxon OX14 4RN

Medicine is an ever-changing science. As new research and clinical experience broaden our knowledge, changes in treatment and drug therapy are required. While many suggestions for drug usages are made herein, the book is intended for educational purposes only, and the author, editor, and publisher do not accept liability in the event of negative consequences incurred as a result of information presented in this book. We do not claim that this information is necessarily accurate by the rigid, scientific standard applied for medical proof, and therefore make no warranty, expressed or implied, with respect to the material herein contained. Therefore the patient is urged to check the product information sheet included in the package of each drug he or she plans to administer to be certain the protocol followed is not in conflict with the manufacturer's inserts. When a discrepancy arises between these inserts and information in this book, the physician is encouraged to use his or her best professional judgment.

Cover design by Marylouise E. Doyle.

Library of Congress Cataloging-in-Publication Data

Hoffmann, Frank W., 1949-
Herbal medicine and botanical medical fads / Frank Hoffmann, Martin Manning.
p. ; cm.
Includes bibliographical references and index.
ISBN 0-7890-1148-4 (hard : alk. paper) — ISBN 0-7890-1149-2 (soft : alk. paper)
1. Herbs—Therapeutic use—Encyclopedias. 2. Fads—Encyclopedias. I. Manning, Martin. II. Title.
[DNLM: 1. Phytotherapy—trends. 2. Plant Preparations. WB 925 H699h 2002]
RM666.H33 H645 2002
615'.321—dc21

2002069099

To our parents,
Frank A. Hoffmann and Lydia Hoffmann
Martin J. Manning and Margaret L. Manning
who have done so much to enrich our lives

And to John Heinerman,
whose works served as an inspiration for this book

ABOUT THE AUTHORS

Frank W. Hoffmann, PhD, MLS, is a professor at the School of Library Science, Sam Houston State University, in Huntsville, Texas. He is the author of *The Development of Collections of Sound Recordings in Libraries, Popular Culture and Libraries, The Literature of Rock* series, and the *Cash Box* chart compilation series. His teaching responsibilities include library collection development, reference/information services, a seminar of popular culture, research methods/grantsmanship, and organization of collections.

Martin J. Manning, MLS, is a research librarian and De Facto Archivist in the Office of International Information Programs, Bureau of Public Diplomacy, United States Department of State, Washington, DC on the Economic Security Team. He was De Facto Archivist for the United States Information Agency from 1978-1999 and is present Co-chair, World's Fairs and International Expositions Section, of the Popular Culture Association/American Culture Association. Manning is a contributor to *American National Biography, The Encyclopedia of New England Culture, Encyclopedia of Gardens: History and Design, Literature of Travel and Exploration: An Encyclopedia,* and authored the chapter on Osaka 1970 in the *Historical Dictionary of World's Fairs and Expositions, 1851-1988.* His areas of research and expertise include popular culture, world's fairs, propaganda, and library history.

CONTENTS

Acknowledgments

Many people and organizations supplied valuable information in the compilation of this volume. Jim Adams, U.S. National Arboretum, Washington, DC, helped lay the groundwork for the book through a series of phone conversations in July 2000. Stanley Cichowicz, Food and Drug Administration, Washington, DC, was invaluable in explaining the FDA regulation of herbs and spices in layman's terms. Marca Woodhams and Valerie Wheat, librarians at the Horticultural Branch Library, Smithsonian Institution Libraries, Washington, DC, functioned as research associates; they assisted in uncovering new sources of data and in putting the authors in touch with valuable contacts. We thank them for their continual efforts and unending courtesies to us in the preparation of this book. Laurie Harrsen, Consumer Products Department, McCormick Spice Company, Sparks, Maryland, defined the types of popularity used to determine trends in spices. Other individuals who provided information and research assistance included Ann Jerabek, Reference and Interlibrary Loan, Gresham Library, Sam Houston State University; Bonnie Boorman, Director, South Branch Public Library, The Woodlands, Texas; Holly Kirkwood, a librarian working in the Houston Independent School District; and Teri Presley, Head of Circulation, Gresham Library, Sam Houston State University.

Organizations and institutions that provided useful information included the National Arboretum Library, a branch of the National Agricultural Library (USDA/SEA/TIS), U.S. Department of Agriculture, Washington, DC; the American Herbal Products Association (AHPA), Silver Spring, MD; and the American Spice Trade Association (ASTA), Englewood Cliffs, NJ.

Informative Web sites included those maintained by the Herb Growing and Marketing Network (HGMN), <www.herbworld.com>; the Herb Research Foundation, <www.herbs.org>; The Herb Society of America, <www.herbsociety.org>; Herbnet, <www.herbnet.com>; and the International Herb Association, <www.iherb.org>.

The authors also owe a debt of gratitude to the countless other friends and associates who provided data and other observations relating in some manner to herbs, spices, and botanicals.

Introduction

This volume is organized in much the same manner as the titles within the Haworth fad encyclopedia series authored by Frank W. Hoffmann and the late William G. Bailey in the 1990s. There is little doubt that herbs, spices, and other botanicals have undergone a huge upsurge in popularity since the 1960s; this is reflected in the present-day sales figures of botanical suppliers and retailers as well as the proliferation of ads and profiles in the mass media. Whether or not any individual plants and associated phenomena (e.g., aromatherapy) have, in fact, achieved fad-like status could be debated at considerable length.

At this point it might be instructive to briefly consider definitions of fads set forth in other sources. In an attempt to integrate a wide range of prior definitions, Hubert G. Blumer offered the following analysis for the revised *International Encyclopedia of the Social Sciences* (1968):

> Fads, like fashion, may occur in widely different areas of group life, such as games, recreation, entertainment, dietary practice, health and medical practice, dress, ornamentation, language, and popular beliefs. . . . Fads have no line of historical continuity; each springs up independent of a predecessor and gives rise to no successor. . . . Fads do not require endorsement by a qualified prestige group in order to gain acceptance; they may spread from any section of hierarchized society. Fads are ephemeral. . . . Fads follow the pattern of a craze or boom, thriving on spectacular and excitatory appearance . . . only to exhaust their attractiveness and undergo a rapid demise. (Volume V, p. 344)

Hoffmann and Bailey attempted to refine faddish behavior to an even greater extent in *Arts & Entertainment Fads* (1990):

> We do not believe that every fad must conform to the skyrocket theory—rapid ascent, rapid fall, then disappearance. Some fads

> have lingered on for years, especially those prior to the frenetic 20th century before mass communications. We do believe that a fad dies out, but not always overnight, and it can reappear at any time in the same or altered manifestation—there is nothing new under the sun. Furthermore, for a fad to be called such it must affect a sizeable number of people and be of enough importance to warrant documentation. We also believe that practically every area of life has been touched by fad, and that within some of those areas evolution has taken place. (pp. xv-xvi)

Herbs, spices, and other botanicals permeate the core fabric of everyday life, whether used for medicinal, therapeutic, cosmetic, ornamental, or dietary purposes. Most people simply regard them as substances that enhance the quality of life. Discerning a notable increase in the popularity of a given plant that has been used for centuries would appear, at first glance, a complicated business. Nevertheless, a concerted effort to plot variations in popularity within the field does pay some degree of dividends.

In the case of spices, consider black pepper. According to researchers in the Consumer Products Department of McCormick and Company, the primary producer of spices globally, this food condiment has been around as a basic seasoning for so long that it has retained a rather predictable popularity. For comparison's sake, consider cumin. Ten years ago, this spice was not well known; today it represents one of the emerging changes in popular taste as it has become one of the most popular spices that McCormick sells.

In her 1990 book, *Herbal Treasures,* Phyliss Shaudys identified a number of fads occurring during the prior two decades. Herb clubs and study groups, an offshoot of herbal associations and societies, owe their upsurge in popularity to the Internet. They are organizing rapidly on the local, state, and national levels, often merging with university and state Cooperative Extension personnel, and offering conferences, seminars, and newsletters to the public. More recently, the promotion of herbs and spices on the Internet has become a full-fledged phenomenon. Given the ease of accessing the Net, community involvement in herb-related associations, societies, and clubs is at an all-time high, and information dissemination has been greatly enhanced by means of e-mail, chat rooms, and the maintenance of personal home pages.

According to Cary Groner in *Herb Quarterly,* two herb-related trends have been evolving since World War II. In North America, a region where the medicinal use of botanicals has lagged behind other parts of the world, an "herbal renaissance" is gathering momentum. Amid a growing awareness of the limitations (e.g., undesirable side effects) of synthetic medications, people are seeking a return to the traditional remedies employed by their ancestors. Herbal research represents the second of these trends. It is most in evidence in the Pacific Northwest; notable examples include the Eclectic Institute, located in Portland, Oregon, or the Williams, Oregon-based Herb Pharm, a combination herb garden, testing and research laboratory, wildcrafting station, organic herb screening and buying center, and tincture producer.

Some fads represent a return to an earlier era. The vogue for Victorian gardens and crafts fit this profile. Today, "Victoriana"—a widely recognized buzzword within gift shops and the decorative arts—encompasses a wide range of botanical applications.

Whatever the respective merits of these fads or trends, it is clearly easier to plot developments on a broader canvas. In other words, nuances in popularity are much harder to discern regarding specific botanicals. This perception is largely reinforced by content analysis of articles cited in the *Reader's Guide to Periodical Literature* (1918 to the present). In his survey of mass circulation magazines, Martin Manning has compiled a time line of herb and spice trends, reflecting to some degree peak levels of popularity in contrast to a more common level of acceptance and use within the fabric of society.

Time Line of Herb and Spice Trends

Pre-1500	Herbals
1500-1900	Bible gardens
	Sachets
	Potpourri, a form of nosegays that were used in the medieval period up to the introduction of better hygiene to disguise bad body odor. Along with aesthetic beauty, there was a practical purpose.
	Herbs and spices in literature
	Library and archival material on herbs and spices

(continued)

(continued)

1922	Condiments
	Herb gardens
1931	Lavender
1936-1950	Yards
1937-	Thyme
1939-	Herb teas
1942-1945	Victory gardens
1945-	Window gardens (boxes, planters)
1968-	Cookbooks
1968-1975	Nasturtium
1971-1980	Mints (plant)
1990-	Internet with chat groups and Web sites (associations, libraries, archives)
	"Brother Cadfael" (TV series). Brother Cadfael, the character created by Ellis Peters, was a monk who used herbs to solve murders. This British TV production did much to popularize herbs and to send viewers back to the Ellis Peters novels that inspired the program.
1995-	Herb gardens
	Growing healthful herbs
1998-	Kitchen spices
1999-	Therapeutic uses of herbs
	Healing herbs

Since World War II, the public fascination with herbs and spices has been heightened by advances in travel and communication (e.g., the Internet) and increased concern for health and the environment. During this period, interest in herbs and spices has also become increasingly diversified. In her *Encyclopedia of Herbs and Their Uses* (1995), Deni Bown notes that as recently as the 1970s, few households had more than a packet of dried mixed herbs. Indian restaurants with their curry-flavored dishes were a luxury enjoyed mostly by the gourmet. An increased interest in cooking with a greater variety of herbs and spices was followed by a resurgence of interest in their use for medicinal and other purposes.

The current interest in herbs follows many diverse paths, from potpourri, dried arrangements, and crafts of all sorts, to aromatherapy, to culinary, landscaping, and medicinal uses. This complexity is evident in the varied applications of individual plants; for example, gingko means considerably different things to a gardener, botanist, and herbalist.

Such varied uses for traditional herbs were unknown to the early herbalists. When Maud Grieve published *A Modern Herbal* (1931), she was responding to a critical shortage of drug imports during World War I. Today, our society deals with the devastation of AIDS and other diseases, the stresses of everyday living, and the high rate of extinction plaguing plant species. When Grieve published her classic work, there was little concern for the natural habitat of herbs and other botanicals. At the outset of the twenty-first century, there exists a very real concern about the propagation of herbs, whether in the wild or through cultivation. For example, the rapid disappearance of South American rain forests is seen as posing a threat to the subsistence of desirable medicinal plants (both those presently in use and others yet to be discovered).

The various uses of herbs and spices throughout history have reflected the practical interests of the period in question. The fields of botany and horticulture originated with the study of herbs as documented in the early herbals. Early works included Dioscorides' *De Materia Medica* (A.D. 512), the oldest illustrated herbal to survive, as well as the *Leech Book of Bald* (tenth century) and *Herbarium* by Apuleius Platonicus (c. 1481). The earliest gardens were herb gardens; these were maintained by monks and nuns during the Middle Ages to guarantee a ready supply of medicinal plants. The Muslim military expansion played a key role in disseminating the knowledge and use of spices in both the Old and New Worlds.

The importance of herbs and spices in everyday life was best reflected by the publication of a steady stream of landmark herbals as reprints. *The Herball; or, Generall Historie of Plantes* (1597), by John Gerard, was a model for many works within the genre during the modern era. It was essentially an index to various plants and medicines, with over 1,300 pages of text and illustrations for the "parfyt knowledge and under standyng of all maner of herbes" (from the herbal's subtitle), and incorporating information from other sources, such as a French herbal, *Le Grant Herbier* (1498). At the time, Gerard was accused of plagiarism when scholars found that the text relied heavily on a translation of Dodoens' *Stirpium Historiae Pemptades Sex* by a member of the London College of Physicians,

Robert Priest, who died before it was first published in 1583. The manuscript allegedly came into the hands of Gerard, who then issued the work as his own. Nevertheless, the book became a "Bible" to medical and garden professionals as well as amateur herbalists who valued its thorough integration of existing knowledge about herbs and somewhat crude drawings.

Many other landmark herbals remain available today. *The Great Herbal of Leonhart Fuchs: De Historia Stirpium Commentarii Insignes* (1542), typifies the current state of affairs, having been reissued as a two-volume boxed set in 1999; it is available in outlets as diverse as the gift shop in the National Museum of Natural History, Smithsonian Institution, Washington, DC, and the American Botanical Council herbal education catalog.

The serious study of herbs and spices has found a place in the curriculum of at least one institution of higher education. In the fall 2001 semester, Bastyr University, Kenmore, Washington, offered a bachelor's degree program in the herbal sciences. The program, requiring ninety credit hours of coursework toward the receipt of an undergraduate diploma, is intended to provide a thorough, scientifically rigorous introduction to the field of herbal medicine. This is considered to be the first accredited herbal sciences degree in the United States, in contrast to the array of self-development courses typically offered in community colleges and local educational centers.

Botanicals are even closely aligned with religious behavior. Many long-standing Christian denominations utilize herbs and spices to symbolize key events in the Church calendar. Wiccan rituals place a special emphasis on the magic qualities of herbs and on the divinity of nature in general. The Web offers a wide range of herb products and literature geared to such practices.

The Internet also abounds with critics of and apologists for less government control of botanicals. The Rocky Mountain Herbal Institute site (www.rmhiherbal.org) typifies the latter approach; in early 2001 it included a link to an extended essay titled, "Stop FDA Attempts to Restrict Availability of Herbs and Natural Products." The article included a survey of recent FDA raids on private citizens, a discussion of concerns over the quality of herbal supplements, recommended strategies for political lobbying, and a lengthy list of source material (print and Web links). Rob McCaleb's Web essay, "Controversial Products in the Natural Foods Market" (www.herbs.

org/greenpapers/controv.html), reflects the conscientious efforts on the part of many apologists to provide rational responses to critics of herb products. McCaleb, the president of the Herb Research Foundation, cites the controversial products noted in a roundtable discussion at New Hope Communication's Winter 2000 Networking Conference and includes a brief defense for each. The listing includes comfrey, chaparral, stevia, ephedra, laxative and stimulant diet teas, hemp, Kombucha "mushroom" products, kava kava, grape seed extract/pine seed extract/Pycnogenol, evening primrose oil/black currant oil, melatonin, tryptophan, chromium picolinate, beta-carotene, single amino acids, germanium, selenium, and organic sugar.

In response to the growing complexity of this field, the present work attempts to integrate information from a variety of perspectives; agriculture and botany, the medical sciences, gardening, the decorative arts, cosmetics, cooking, and popular culture. In addition to a basic description of herbs and spices, augmented by discussions of their cultivation and use, the book looks at their history and folklore, governmental regulation, notable clinical research, and popularity with the general public. Geared toward both herb enthusiasts and students in schools and institutes of higher education, the entries encompass notable herbs and spices, industry organizations, social customs, and consumer commodities. Limitations of space dictated the omission of a considerable number of lesser-known, albeit culturally important, botanicals as separate entries. An inventory of these plants is provided in the Appendix, along with a brief listing of their respective uses. For further information, readers are advised to consult the works cited in the Recommended Reading list at the end of the book.

BIBLIOGRAPHY

Bown, Deni. *Encyclopedia of Herbs and Their Uses.* London and New York: Dorling Kindersley, 1995.

Grieve, Maud, and Leyel, Hilda (Eds.). *A Modern Herbal.* 2 vols. New York: Dover, 1981.

Groner, Cary. "Herbal Innovators." *Herb Quarterly 40* (Winter 1988-1989): pp. 7-9.

Hoffmann, Frank, and Bailey, W. *Arts and Entertainment Fads.* Binghamton, NY: The Haworth Press, 1990.

McCaleb, Rob. "Controversial Products in the Natural Foods Market." www.herbs.org/greenpapers/controv.html/, 1997.

Shaudys, Phyliss. *Herbal Treasures.* Pownal, VT: Storey Communications, 1990.

Sills, David L. (Ed.). *International Encyclopedia of the Social Sciences.* 19 vols. New York: Macmillan [1968], 1991.

A

ALFALFA

Alfalfa *(Medicago sativa)* is a perennial herb typically found rimming fields in low-lying valleys. It consists of a smooth stem that grows from an elongated taproot and reaches a height of a foot or more. Blue to purple flowers appear during the summer months, eventually producing spirally coiled seed pods.

The ancient Persians fed their horses alfalfa to make them look sleeker and feel stronger. The Arabs designated it "the father of all food," employing it as livestock feed. It continues to be widely used in this capacity by modern farmers due to its high nutritional value. The herb includes vitamins A, B-1, B-6, B-12, C, E, K-1, niacin, biotin, folic acid, pantothenic acid, and others. It also contains many amino acids, proteins, naturally occurring sugars (e.g., sucrose, fructose), major minerals, and trace elements such as iron, calcium, magnesium, potassium, phosphorus, zinc, and copper.

Because of their exceptional nutritional value, alfalfa sprouts make an excellent lettuce substitute. According to the August 1984 issue of the *Journal of Nutrition* (cited by Heinerman, 1996, p. 9), scientists at the University of California at Davis determined that the high manganese content in alfalfa extracts improved the condition of diabetics who failed to respond to insulin.

The herb has many other medicinal applications as well. Clinical nutritionists have demonstrated that alfalfa meal, when eaten by lab monkeys whose diets included considerable amounts of cholesterol, helped prevent atherosclerosis in addition to reducing serum cholesterol levels. Dr. Henry G. Bieler included the following story in his best-selling book, *Food Is Your Best Medicine,* to illustrate alfalfa's value as an infection fighter. Coming upon a farmer in rural Idaho who was suffering from a very bad leg ulcer (to the point that the entire limb seemed close to becoming gangrenous), Bieler recommended a blend of chopped alfalfa shoots and equal parts water and canned grapefruit juice. The leg condition eventually healed completely. The plant's rich chlorophyll content has been used by doctors in some major hospitals since the 1940s to treat infections arising from surgical incisions, bedsores, and inner ear conditions. In these situations, patients drank fresh juice made by running raw alfalfa sprouts through a blender. Applying the juice externally to surface infections has also proven effective.

A Mormon farmer has recounted the following application of the herb in his autobiography, *History of a Pioneer: Edward Leo Lyman, Jr. (1881-1958):*

> . . . [my father] said to me, "If you ever get the blues a sure cure is to go out and look at your alfalfa a while." I've recommended this bit of sage advice to several farmers who faced foreclosures and were in deep depressions. A couple of them followed it and claimed afterwards just how therapeutic it was to gaze out on their fields of uncut alfalfa hay for an hour or so.

BIBLIOGRAPHY

Bieler, Henry G. *Food Is Your Best Medicine.* New York: Vintage Books, 1973.

Heinerman, John. "Alfalfa *(Medicago sativa).*" In: *Heinerman's Encyclopedia of Healing Herbs & Spices.* Paramus, NJ: Prentice-Hall, 1996, pp. 7-10.

Lyman, Melvin A. *History of a Pioneer: Edward Leo Lyman, Jr. (1881-1958).* Delta, UT: M.A. Lyman, 1959.

ALLSPICE

Allspice *(Pimenta dioica)* is derived from a tall tree native to Latin America and the Caribbean. Spanish explorers discovered it shortly after Columbus came upon the islands. Because the dried berries resembled black peppercorns in shape and color, they were referred to as "pepper." The Aztecs employed allspice to sweeten and flavor their favored chocolate drink. The spice was exported to Europe from 1601 onward as a substitute for cardamom. During the seventeenth century, pirates utilized it to smoke and barbecue meat prior to their expeditions in West Indian waters. The pirates called the wooden grid used to smoke the meat boucan after the French verb "boucaner" for cured or barbecued; as a result, they became known as buccaneers.

Possessing a slim trunk that sheds its soft, light-grey bark annually, the allspice tree branches high above the ground, bearing pairs of lanceolate leaves. The leaves—shiny dark-green on top and lighter underneath—possess the same aromatic properties as the berries (most notably, eugenol).

Although best known as a culinary accent, allspice is also popular in the Caribbean as an analgesic for toothaches. The berries are mashed while still green and left to dry in the sun; the resulting powder is then pounded further until it attains a fine consistency. John Heinerman relates that a Creole-speaking herbalist based in Montego Bay kept this powder in a chewing tobacco can. Whenever a toothache occurred, the herbalist dipped a wet forefinger into the powder and rubbed it inside his mouth along the gum line, relieving the pain almost immediately.

Heinerman also tells of an executive secretary who used allspice powder to mask bad mouth odor. The treatment consists of mixing the powder in a glass of warm water; the liquid is then swished around in the mouth cavity for a minute or so.

The majority of the world's supply of allspice comes from Jamaica. Honduras, Guatamala, the Leeward Islands, and Mexico also produce the spice, though the quality is inferior.

BIBLIOGRAPHY

Farrell, Kenneth T. "Allspice." In: *Spices, Condiments, and Seasonings*. Westport, CT: AVI, 1985, pp. 26-30.

Heinerman, John. "Allspice *(Pimenta dioica)*." In: *Heinerman's Encyclopedia of Healing Herbs & Spices*. Paramus, NJ: Prentice-Hall, 1996, pp. 10-11.

ALOE

Aloe *(Aloe vera)* is a perennial succulent native to East and South Africa; it is presently cultivated throughout the tropics, most notably in the West Indies. Aloe's fibrous root produces a rosette of fleshy basal leaves. The leaf tissue contains a mucilaginous extract from which aloe gel is produced.

Aloe possesses an illustrious past. It was allegedly used to embalm Pharaoh Ramses II as well as to preserve the physical remains of Jesus Christ. Immediately prior to setting sail for the New World, Columbus wrote in his diary, "All is well, aloe is on board!"

John Heinerman states that no other herb has been touted so highly by medical and dental practitioners in modern times. Aloe contains ninety-six percent water, making it possible to supply that vital ingredient to injured tissue without closing off the needed air supply. According to the December 1981 issue of *Runner's World* (Heinerman, 1996), the balance of the plant "contains complex carbohydrate molecules, believed essential to aloe's natural value as a moisturizer. Substances present include . . . enzymes, trace sugars, a protein containing 18 amino acids; vitamins; minerals like sulphur, silicon, iron, calcium, copper, sodium, potassium, manganese, and more. The mixture of active ingredients in aloe is called aloin . . . It's responsible for the plant's healing properties" (p. 12).

Aloe has a proven track record in the treatment of a wide range of conditions, including:

- X-ray burns, sunburn, and chemical burns
- Traumatized tissue (after normal and regular cleansing)
- Decubitus ulcers or bedsores
- Primary candidal dermatitis (skin inflammation caused by infection of the yeast *Candida albicans*)
- Stomal ulcers (intestinal ulcers between the stomach and the jejunum)
- Herpes simplex
- Periodontal surgery

- Plant stings, insect bites and stings, and other minor dermatological manifestations

Furthermore, a Dallas, Texas, oral surgeon reported outstanding success in using aloe to treat facial edema, immediate denture placement, lockjaw, and mouth sores. The herb is available medicinally as a prepared ointment, salve or lotion, liquid drink concentrate, and encapsulated powder.

Aloe's reputation as miracle cure-all is reinforced by one Lubbock, Texas, woman's story. Possessing a wart on her arm as large as a pencil eraser, she began covering it with a piece of cotton soaked in aloe gel. Additional gel was applied every three hours. The wart had disappeared after two weeks without the slightest trace of a scar.

BIBLIOGRAPHY

Griggs, Barbara. *Green Pharmacy: The History and Evolution of Western Herbal Medicine.* Rochester, VT: Healing Arts, 1997.

Heinerman, John. "Aloe *(Aloe vera).*" In: *Heinerman's Encyclopedia of Healing Herbs & Spices.* Paramus, NJ: Prentice-Hall, 1996, pp. 11-12.

Hollis, Sarah. "Aloe." In: *The Country Diary Herbal.* New York: Henry Holt, 1990, p. 30.

ALTERNATIVE MEDICINE

Alternative medicine has enjoyed a rapid upsurge in popularity since the 1960s when many of today's major social movements—including Eastern spirituality, ecology, and the back-to-nature ethic—had their genesis. The Discount Natural Herbs home page states that by 1998 60 percent of all U.S. citizens were using some form of alternative medicine.

Alternative medicine is perhaps best described as people taking an active role in their own health care programs, rather than relying on the prescriptive approach employed by many medical practitioners today. It attempts to determine the reason behind a particular medical condition; in contrast, mainstream physicians often try to simply mask an illness by means of a medication or treatment that is currently in vogue with the health care establishment. It encompasses many therapies, most notably acupuncture, Ayurvedic and Traditional Chinese Medicine, herb and vitamin use, chiropractic, homeopathy, naturopathy, massage treatment, and reflexology.

BIBLIOGRAPHY

Griggs, Barbara. *Green Pharmacy: The History and Evolution of Western Herbal Medicine.* Rochester, VT: Healing Arts, 1997.

"Why Choose Alternative Medicine?" Discount Natural Herbs [Home Page]. www.discountnaturalherbs.com. Lacey, Washington, c2001. 2p.

AMERICAN BOTANICAL COUNCIL

Founded in 1989, the American Botanical Council (ABC) serves as a clearinghouse for scientific information on medicinal herbs. Its mission consists of educating the media, government agencies, health care practitioners, and the public on the safe and effective use of medicinal plants and phytomedicines. A renovated farmhouse, built during the 1850s by one of the first millers to settle in Austin, Texas, serves as headquarters. ABC's two-and-a-half acre site is landscaped with medicinal gardens and includes a state-of-the-art greenhouse and rainwater collection system.

Effective winter 2001, ABC instituted a charter membership plan that offered the following benefits:

- Four issues of *HerbalGram*. Nominated as one of the best alternative magazines by *Utne Reader* for two of the past three years, *HerbalGram* is recognized as one of the leading sources of information on herbal research.
- A collection of back issues of *HerbalGram*.

- Unlimited use of an extensive members-only section of the ABC Web site: www.herbalgram.org. Honored by Microsoft for its user-friendliness in 2001, it is a cornucopia of information and resources.
- Discounts on products and publications available through ABC's Herbal Education Catalog.
- Access to HerbClip, ABC's clipping service, which provides 1,200 critical reviews of seminal research articles from 300 publications on CD-ROM, at www.herbalgram.org, or through full-service mailing.
- Invitations to book signings and special events at ABC's Case Mill Homestead. The fifteen medicinal plant gardens at the Austin headquarters can be visited during regular business hours.
- First-notice invitations to participate in once-in-a-lifetime ethnobotanical tours and educational opportunities. Continuing education credits are provided for health care professionals.

The Council's future plans include continuing education classes for health care providers, herb growers, and the public at large; more book signings, book and plant sales, and medicinal theme gardens; expanded tours of the gardens conducted by visiting herbalists; an upgraded resource center; and additional interaction with the local community.

ABC is accessible at P.O. Box 144345, Austin, Texas 78714-4345; phone: (512) 926-4900; fax: (512) 926-2345; e-mail: abc@herbalgram.org; Web site: www.herbalgram.org.

BIBLIOGRAPHY

American Botanical Council Home Page: www.herbalgram.com.

Blumenthal, Mark. Cover letter—with enclosures—from founder and executive director of the American Botanical Council, announcing charter membership program, Winter 2001.

HerbalGram. Quarterly. Austin: American Botanical Council, 1984-. [individual issues].

AMERICAN HERBAL PRODUCTS ASSOCIATION

The American Herbal Products Association (AHPA) was founded in 1982 as the national trade association and voice of the herbal products industry. It is composed of companies doing business as growers, importers, manufacturers, and marketers of herbs and herbal products. The AHPA serves its members by promoting the responsible commerce of products that contain herbs. Most AHPA members sell botanicals as dietary supplements or as over-the-counter (OTC) monograph drug products. Some are also interested in the development of new botanical drug products.

To ensure that herbal manufacturers provide material information about their products, the AHPA has developed specific labeling guidelines for a number of botanical ingredients. Labeling recommendations exist for products containing chaparral, comfrey, ephedra, kava, saw palmetto; and St.-John's-wort, among others. In addition, the AHPA published the *Botanical Safety Handbook* (1997), which classifies over 500 herbs within safety categories that can assist both manufacturers in labeling and consumers in making informed choices in their use of herbs.

All of the AHPA's members agree to abide by a Code of Ethics that requires adherence not only to established regulations, but also to meaningful industry policies. Thus, all member companies are expected to follow certain business practices, regardless of whether they are mandated by any government agency. One such measure, established as an industry regulation in 1992, called upon all of the AHPA's members to agree to a single standardized common name for each of the herbs used in their products to ensure clear labeling for consumers. This policy has now been adopted as federal law. Links to many of the AHPA's members and a copy of the Association's Code of Ethics can be found at www.ahpa.org. The Association generally recommends that the consumer buy herbal products from a reputable company.

One area with a different regulatory policy represents an acknowledgment that botanical drugs are usually prepared as complex mixtures. The Food and Drug Administration (FDA) refrains from treating botanical drug products derived from a single part of one species as combination drugs; it further states its intention to propose revisions that will allow similar exemptions under certain circumstances for botanical drugs derived from multiple parts of one species, or from parts of different plant species. The AHPA agrees that such exemptions should be made so that most botanical drugs will not be treated as combination drugs.

At the same time, AHPA believes that accurate identification of all over-the-counter drug substances, including botanical drug substances, should be assured. In the case of botanical drug substances, this can be accomplished by choosing specific methods and references that are relevant to the particular species as well as to the form of the plant at the time that the botanical raw material comes into possession of a manufacturer, e.g., whole, cut, powdered, extracted.

Herbs of Commerce

AHPA's *Herbs of Commerce,* now in its second edition, was adopted as the FDA's standard reference for naming botanical ingredients on dietary supplement labels. According to Michael McGuffin, AHPA President, the Association anticipates that the "FDA will accept the *Herbs of Commerce* revision as authoritative, thus making it the standard reference labeling for hundreds more therapeutic botanicals on the market."

The new edition has been expanded from the 550 plants originally listed to include over 2,000 plants in trade. Its primary editors, Drs. John Kartesz, Arthur Tucker, and Albert Leung, have overseen review of the nomenclature. Two new useful alphabetical lists have been added, one by Latin binomial and the other by standard common name. In addition, qualified experts have provided Chinese (pinyin) names, among other languages, for hundreds of the listed plants. The FDA is currently reviewing the revisions before deciding to adopt the second edition of the book as a new standard for dietary supplement labeling.

AHPA first published *Herbs of Commerce* in 1992 as an attempt to produce a self-regulatory model to address consumer confusion about the common names of botanical ingredients. In their final rules adopted to implement the Dietary Supplement Health and Education Act (1994), the FDA determined that "[t]he common or usual name of ingredients of dietary supplements that are botanicals (including fungi and algae) shall be consistent with the names standardized in *Herbs of Commerce*." The rules further state that the Latin binomial name of plants would be required, "except that this name is not required when it is available in the reference entitled *Herbs of Commerce* for the common or usual name listed on the label."

Presently, the American Herbal Products Association is located at 8484 Georgia Avenue, Suite 370, Silver Spring, MD 20910; phone: (301) 588-1171; fax: (301) 588-1174; e-mail: ahpa@ahpa.org; Web site: www.ahpa.org.

BIBLIOGRAPHY

American Herbal Products Association Home Page: www.ahpa.org.

Bennett, J., and Brown, C.M. "Use of Herbal Remedies by Patients in a Health Maintenance Organization." *Journal of the American Pharmaceutical Association 40*(3) (2000): 353-358.

Tyler, Varro E. *Herbs of Choice*. Binghamton, NY: The Haworth Press, Inc., 1994.

AMERICAN SPICE TRADE ASSOCIATION

Founded in 1907, the American Spice Trade Association is—despite its name—truly global in orientation, serving members in more than thirty-four spice-producing nations. Its family includes:

- Active members (agents, blenders, brokers, dehydrators, distributors, extractors, growers, importers, processors, and traders)
- Associate members (customhouse brokers, laboratories, manufacturers, samplers/weighers, sterilizers, and suppliers)
- Allied members (users of spices in finished food products)
- International members (previous entities located outside the United States)

According to the ASTA home page its objectives are as follows:

- To protect the interests and promote the welfare of the industry
- To assume as a group those functions which the individual member firm cannot perform as effectively
- To encourage and achieve greater international cooperation, understanding, and knowledge leading to continued growth in spice consumption

ASTA activities span a wide range of fields, including:

- Government relations; more specifically, maintaining a dialogue and interfacing with government agencies that relate to spices under the umbrella and effectiveness of group action, monitoring all regulatory and legislative matters that bear on our products and industry, and keeping members updated on these matters
- Public relations (e.g., monitoring and managing issues of potential industry concern, including crisis response)
- Technical services (e.g., pesticide monitoring, sterilization alternatives)
- Trading services
- Information services (e.g., trade statistics, publications, knowledge bank, and data management)
- Business networking opportunities (e.g., annual convention and regional meetings)
- Educational seminars
- Global communications

The publication of *The Jungle* (1906), by Upton Sinclair (one of the authors President Theodore Roosevelt accused of being a muckraker), indirectly affected the way the spice industry conducted business. Although the novel itself had nothing to do with spices, it gave a scathing portrayal of the disgracefully unsanitary conditions of the Chicago stockyards. It caused such a furor in the country about the safety of the nation's food supply that Congress passed two specific acts concerning food safety—the Meat Inspection Act and the Pure Food and Drug Act—the same year.

Against this background, a meeting to form an association of spice traders was held on August 7, 1907, in the rooms of the Underwriters' Club in New York City, with John Clarke as temporary chairman. Its purpose was to act on the new food legislation or be forced out of business. From these discussions, one of the participants proposed that a new organization, the American Spice Trade Association, be established with headquarters in New York City, and that it be incorporated under New York State law. Importers, dealers, agents, manufacturers, and brokers were eligible for membership, with each firm having a single vote.

In 1927, Congress created the Food, Drug and Insecticide Administration; four years later, it dropped "Insecticide" from its name. Before then, food legislation was administered by a Board of Food and Drug Inspection. With the onset of the Great Depression in 1929, the spice business began to suffer against a backdrop of declining markets, poor publicity, intraindustry strife, and a lack of government standards at the import level. ASTA attempted to reorganize; a professional firm, A. H. Doolittle Co., was called for advice. Doolittle noted that a Grinders and Processors section had been written into the bylaws in 1922 and recommended the development of additional sections. The firm also suggested forming additional committees and hiring a paid executive to run the day-to-day operations of the office. The reorganization lasted until 1974 when a second restructuring took place.

Historically, the Association had as its nucleus the trading aspect of the spice business; the various trading contracts and an arbitration system were firmly established. The structure of the Association remained oriented to these activities over time. In 1972, the Board observed inconsistency in ASTA's organization and operation in relation to the interests and demands of government, industry, and consumers. The issues of concern included quality control, sanitation, microbiology, product liability, standardization of product specifications, packaging and labeling requirements, and nutrition. None of these issues were a problem at the time of ASTA's founding but as this changed, many firms joined ASTA for collective strength and to protect their interests.

ASTA is now challenged, in a period of global marketing, instantaneous communication, and rapid social and technological change, to reinvent itself again, to preserve what it has achieved and to set an agenda for the twenty-first century. Whatever changes ASTA makes, it will try to adhere strictly to Clarke's 1907 declaration that ASTA would uphold the law in every possible way and not try, in any case, to oppose, evade, or attempt to discredit it.

ASTA can be reached at 560 Sylvan Avenue, P.O. Box 1267, Englewood Cliffs, NJ; phone: (201) 568-2163; fax: 201-568-7318; e-mail: info@astaspice.org.

BIBLIOGRAPHY

"About the American Spice Trade Association." www.astaspice.org. Englewood Cliffs, NJ: American Spice Trade Association, n.d. 4p.

Garland, Sarah. *The Complete Book of Herbs and Spices*. Pleasantville, NY: Reader's Digest Association, 1993.

Swahn, Jan O. *The Lore of Spices: Their History and Uses Around the World*. New York: Barnes and Noble Books, 1997, c1991.

ANEMONE

A member of the buttercup family, anemone *(Anemone pulsatilla)* reaches five inches to a foot in height. Rather than petals, it possesses eight to ten bright sepals ranging from cream to lavender-red in color. The floral portions include numerous yellow stamens, which culminate in a conical, silky seed head. The leaves approximate those of parsley, but often include purple highlights.

Anemone has a long history of medicinal application. Once widely associated with the treatment of leprosy, the plant is now used to help heal skin lesions on AIDS patients. The concoction consists of simmering the leaves in distilled water; the liquid by-product is strained into a glass container, to be applied twice daily in the form of a sponge bath. While in New York City to give a health lecture at a New Age expo, John Heinerman was approached by a man concerned about AIDS-related lesions that covered his body. After trying the anemone tea preparation, the AIDS patient reported that 75 percent of the sores were healed.

BIBLIOGRAPHY

Bradley, P.R. (Ed.). *British Herbal Compendium.* Volume I., London: British Herbal Medicine Association, 1992.

Heinerman, John. "Anemone *(Anemone pulsatilla).*" In: *Heinerman's Encyclopedia of Healing Herbs & Spices.* Paramus, NJ: Prentice-Hall, 1996, pp. 15-16.

Hoffmann, D. *The New Holistic Herbal.* Dorset, U.K.: Element, 1990.

Mills, S. *The Complete Guide to Modern Herbalism.* London: Thorsons, 1994.

ANISE

An annual of the family *Apiaceae,* anise *(Pimpinella anisum)* is related to other plants bearing aromatic fruit (commonly referred to as seeds), including dill, cumin, caraway, and fennel. Cultivated plants range from eighteen inches to four feet in height, accented by finely divided, feathery, bright-green leaflets. The name *Pimpinella* (from the Latin *dipinella*) is derived from the twice-pinnate form of the leaves. White flowers up to three inches across appear in the umbels as early as six weeks after sowing; the seeds ripen to a greyish-green color in approximately four months. The entire plant is imbued with a pronounced anise taste.

Near Eastern in origin, anise has had medicinal and fragrance uses since antiquity; it was allegedly popular in Egypt as early as 1500 B.C. Pliny the Elder claimed that if it were tucked under one's pillow at night, inhalation of the aroma would prevent disagreeable dreams. Valued as an aphrodisiac, the seeds have also been employed to flavor cookies, cakes, and liqueurs (e.g., anisette) as well as to season fatty meats. The latter practice, and the Roman tradition of concluding feasts with anise seed cakes, probably originated as a digestive aid. In the sixth century, Pythagoras believed that the herb was useful for scorpion bites, while others wrote that it promoted the flow of urine, sweetened the breath, and prevented shortness of breath. In addition, the oil from the seeds has been used to treat digestive problems and bronchial conditions, to fight lice, and as a rodent bait. Its popularity was such that the Virginia Assembly decreed in 1619 that each household should plant at least six anise seeds and repeat the plantings on an annual basis.

Anise is frequently confused with other plants due to similarities in taste and aroma. Star anise *(Illicium verum)*—the fruit of a tree flourishing in China and Japan—bears a seed in each section of its star-shaped arrangement. Florence fennel or finocchio (the vegetable form of fennel) is often labeled anise in grocery stores.

BIBLIOGRAPHY

Farrell, Kenneth T. "Anise Seed." In: *Spices, Condiments, and Seasonings*. Westport, CT: AVI, 1985, pp. 31-35.

Foster, Gertrude B., and Louden, Rosemary F. *Park's Success with Herbs*. Greenwood, SC: G.W. Park Seed Co., 1980.

Herb Society of America. "An Herb Society of America Fact Sheet: Anise," http://www.herbsociety.org/anise.htm. April 1999.

Reader's Digest Association, Inc. *Magic and Medicine of Plants*. Pleasantville, NY: Reader's Digest, 1986.

Wetherilt, W. and Pala, M. "Anise (Anason)." In Charalambous, George (Ed.), *Spices, Herbs and Edible Fungi*. Amsterdam: Elsevier, 1994, pp. 290-292.

APHRODISIACS

Named for Aphrodite, the Greek goddess of love and beauty, an aphrodisiac represents any substance or activity—as perceived by the five senses—that stimulates sexual desire. Reactions triggered by its use can vary widely, from inducing physical changes in the body to enhancing a mood by sensual associations or by spurring the recollection of pleasant erotic memories.

Aphrodisiacs fall into three general categories: substances whose sexual appeal has strong erotic associations; those enhancing physical health and peace of mind, thereby enabling the mind and body to explore erotic connections; and those having a physical or erotic effect in and of themselves.

The first group includes many romantic traditions: flowers, perfume, moonlight, exotic locations, etc. Herbs have played an important role for centuries in helping stimulate erotic attraction. Herbal cosmetics, hair rinses, and cleansers all figure prominently in erotic stories and myths. Notable herbs here include myrtle, rosemary, thyme, mint, and lavender.

Herbs reputed to enhance sexual health include celery, cloves, galangal, ginger, juniper, Lady's Mantle, laurel, nasturtium, and parsley carnations. Ginseng and garlic have probably garnered the greatest attention in recent decades; the latter requires that both sexual participants ingest it in order to ensure mutual compatibility.

A scientific basis may exist for claims relating to the erotic effect of herbs on the human body. Robbie Cranch notes that "a number of herbs recommended as aphrodisiacs contain naturally occurring hormones that have been synthetically replicated and used in forms of hormone therapy and birth control. For example, licorice root and wild sarsaparilla, two herbs widely used historically as aphrodisiacs . . . have served as analogues for the chemically derived versions of estrogen and testosterone" (p. 58).

The physical aphrodisiac effects of certain herbs may be a product of the strong belief that they work. Parsley and thyme both possess a rich tradition in the enhancement of lovemaking. Calypso, the seductress of Greek antiquity, lived on an island covered with parsley. Victorian girls were admonished not to cut parsley, for that would make them unlucky in love, nor give it away, for to do so would give away their luck in love. On Saint Agnes' Eve (January 20), a questing woman would blend thyme with rosemary while offering the prayer, "Saint Agnes, that's to lovers kind, Come, ease the trouble of my mind." In return, the virgin martyr would reputedly send a dream about the woman's true love.

Herbal aphrodisiacs are documented to have played many other roles in the courtship ritual, including divination and the use of symbolic plants as gifts. Lovers have long exchanged tussie mussies including herbs and flowers such as arborvitae (representing unchanging friendship), carnation (admiration), celandine (joys to come), daffodil (regard), fennel (strength), ivy or fir (fidelity), lavender (loyalty), mistletoe ("I surmount difficulties"), olive (peace), rose (love), rosemary (remembrance), sage (domestic virtues), tulip (ardent love), and violet (steadfastness).

While the verdict is inconclusive regarding whether or not herbal aphrodisiacs really work, many would argue that the positive associations they bring to an emotional and physical relationship (e.g., healing, sensual delight, thoughtfulness) render them a valuable tool in lovemaking.

BIBLIOGRAPHY

Bavard, Tania. *Sweet Herbs and Sundry Flowers*. Boston: David R. Godine, 1985.
Cranch, Robbie. "Herbs of Love." *The Herb Companion*. (October/November 1995): 56-59.
Douglas, Norman. *Venus in the Kitchen*. London: Michelin House, 1952. Reprinted in 1992.
Lehner, Ernest, and Lehner, Johanna. *Folklore and Symbolism of Flowers, Plants and Trees*. New York: Tudor, 1960.
Raetsch, Christian. *The Dictionary of Sacred and Magical Plants*. Dorset, U.K.: Prism Press, 1992.
Weddeck, Harry E. *A Dictionary of Aphrodisiacs*. New York: M. Evans, 1989.

ARNICA

An aromatic perennial, arnica *(Arnica montana)* can be identified by its creeping rhizome, which produces a basal rosette of four to eight downy leaves. The flowering stem—generally unbranched—is hairy, one to two feet in height, and possesses just one or two pairs of opposite leaves. The golden yellow flowers (reminiscent of daisies) appear from mid-summer to early autumn. The plant is most prevalent in the central and northern regions of the northern hemisphere, most notably locales featuring sandy acid soils, ample humus, and plenty of sun.

The famous German poet Johann Wolfgang von Goethe (1749-1832) spoke highly of the merits of arnica tea, brewed from the plant's flowers. He claimed it was always effective in eliminating the angina pain arising from hardening of his heart arteries (now known as arteriosclerosis).

John Heinerman relates that a famous German opera star touted arnica tea for clearing up her sore throat and laryngitis. Just a few minutes after gargling with it, her throat felt as good as new, enabling her to go back on stage and continue singing.

Arnica is typically the first choice of homeopathic and naturopathic practitioners in treating sports injuries such as severe bruises, sprains, wounds, and shock. It is also popular as a remedy for diabetic leg ulcers. Treatment for such conditions involves applying the herb as a tincture, ointment, or oil (but only where the skin is unbroken).

BIBLIOGRAPHY

Garland, Sarah. *The Complete Book of Herbs & Spices*. Pleasantville, NY: Reader's Digest Association, 1993.

Heinerman, John. "Arnica *(Arnica montana)*." In: *Heinerman's Encyclopedia of Healing Herbs & Spices*. Paramus, NJ: Prentice-Hall, 1996, pp. 23-25.

AROMATHERAPY

Aromatherapy involves the utilization of essential oils to cure physical and psychological ailments as well as enhance general well-being. The activity was well established in ancient cultures worldwide. Medical specialists in the powerful dynasties of both China and Egypt made it a cornerstone of their practice. The ancient Greeks documented the mood-altering properties of a blend of oils known as kyphi.

In aromatherapy, the oils can be administered through the skin, by way of massage; through the olfactory passage, by inhalations; or through the digestive tract, by means of tisanes. The absorbed oils are then taken up selectively by different tissues, producing a soothing effect as well as reinforcing the body's own natural healing processes.

Because the oils are highly concentrated and possess diversified properties, they should be diluted in a base or carrier oil. A maximum of five drops of an essential oil to one teaspoon of carrier oil is generally recommended by experts. The latter may be a single oil or a mixture; popular choices include grapeseed, peanut, hazelnut, walnut, soybean, and almond oils.

Given the current popularity of aromatherapy, a wide range of essential oils are currently available in the marketplace. They include:

- Bergamot—helps lift depression
- Chamomile—relieves anxiety and dispels anger
- Coriander—combats tiredness and lethargy
- Frankincense—produces a calming effect; particularly effective when used in a vaporizer
- Geranium—soothes and relaxes
- Grapefruit—warming and soothing; functions best as a confidence booster
- Jasmine—soothes anxiety and helps bring on relaxation and sleep
- Lavender—acts as a relaxant and steadies the nerves
- Melissa—helps avert depression and induce a sense of well-being
- Neroli—eases nervous tension and anxiety and encourages sleep
- Rose (or attar of roses)—also eases tension and anxiety, particular with respect to postnatal depression
- Rosemary—aids in overcoming exhaustion, stimulating the senses, and enhancing concentration
- Sandalwood—reduces anxiety and nervous tension
- Ylang-Ylang—functions as a sedative and antidepressant

The blending of essential oils results in synergistic compounds, each of which is recognized as having unique properties and actions. Less experienced users are advised to seek guidance from knowledgeable practitioners prior to the blending process.

BIBLIOGRAPHY

Gardner, Jo Ann. "Aromatherapy." In: *Living with Herbs*. Woodstock, VT: Countryman Press, 1997, pp. 61-62.

Green, Mindy. *Natural Perfumes: Simple Aromatherapy Recipe.* Loveland, OH: Interweave, 1999.

Green, Mindy. "Ritual Oils." *The Herb Companion Corner.* 2p. Reprinted from February/March 2000 issue of *The Herb Companion.* www.discoverherbs.com/herbcompanion.

Rose, Jeanne. *The Aromatherapy Book.* Berkeley, CA: North Atlantic Books, 1992.

Tisserand, Robert. *Aromatherapy.* Twin Lakes, WI: Lotus Light Press, 1988. Also published as *Aromatherapy for Everyone* (New York: Penguin, 1988).

ASHWAGANDHA

The increasing awareness of ashwagandha *(Withania somnifera)* in North America during the 1990s coincided with the upsurge in popularity of the ancient healing art from India known as Ayurveda. Ayurveda's success owed much to a series of best-selling books by Dr. Deepak Chopra, most notably, *Quantum Healing* (Bantam, 1989) and *Perfect Health* (1990). According to Chopra, "Ayurveda is nothing more than the human mind exerting its deepest influence on the body, thereby making us aware of the need to bring it more into balance with nature" (p. 35). Chopra believes that various herbs and spices, including ashwagandha, help facilitate the optimum balance between mind and body; the result is freedom from sickness.

The name refers to the plant's odor, described as smelling much like a sweaty horse. An erect shrub possessing greenish or bright-yellow flowers, ashwagandha reaches seven feet in height in India, Pakistan, the Middle East, and the eastern Mediterranean. Its active ingredients—alkaloids and withanoloids—are similar to those in ginseng; in fact, both herbs are utilized to enhance longevity and sexual performance. Whereas ginseng is a stimulant, however, ashwagandha acts as a tranquilizer and sedative. It is as esteemed in the Middle East as ginseng is in the Orient.

Ashwagandha has gained considerable support as an effective medicinal within mental health circles. A study outlined in *Selected Medicinal Plants of India* (1992) indicated that patients taking regular doses of the herb for one month exhibited little evidence of prior anxiety disorders, panic attacks, and similar mood disorders. Likewise,

American psychiatrists have determined that psychotic disorders such as manic depression, alcoholic paranoia, and schizophrenia have been effectively treated by its use. Furthermore, a combined regimen of ashwagandha, gotu kola, and ginkgo biloba capsules has been proven to dramatically enhance learning ability and memory retention.

Studies on both rats and humans have conclusively shown that the herb (in alcoholic root extract form) has helped put cancer tumors into regression and reduce inflammation in rheumatoid arthritis. Ashwagandha's high steroid content has proven more potent than hydrocortisone for the latter condition. It is also being used to treat sports injuries.

The herb appears to possess many other potential benefits. It kills several types of harmful bacteria and fungi. Animal research has revealed that ashwagandha lowers blood pressure and protects the liver from damage induced by toxic chemicals.

BIBLIOGRAPHY

Castleman, Michael. "Ashwagandha: Hard to Say, Good to Know." *The Herb Quarterly* (Summer 2000): 6.

Chopra, Deepak. *Perfect Health.* New York. Harmony, 1990.

Chopra, Deepak. *Quantum Healing.* New York: Bantam, 1989.

Heinerman, John. "Ashwagandha *(Withania somnifera).*" In: *Heinerman's Encyclopedia of Healing Herbs & Spices.* Paramus, NJ: Prentice-Hall, 1996, pp. 35-36.

Selected Medicinal Plants of India. Bombay: Chemexerl, 1992.

B

BABY'S BREATH

Possessing one of the most poetic names in the plant kingdom, baby's breath *(Gypsophila paniculata)* has a long history of popularity with florists as a counterpoint to roses, carnations, and the like in floral arrangements. It is easily recognizable by the spray of tiny white, pink, or reddish flowers appearing on a profuse web of small branches. Related perennials within *Gypsophila* are widely used in outdoor gardens.

While best known for its decorative uses, baby's breath has also proven effective in the treatment of bronchitis. An Austrian physician specializing in the treatment of respiratory disorders, Dr. Ahron Forschung, advocates use of a hot tea brewed from the plant's roots. Drinking a cup of the beverage every five hours or in between meals helps dislodge the mucus and facilitate better breathing.

BIBLIOGRAPHY

Heinerman, John. "Baby's Breath *(Gypsophila paniculata)*." In: *Heinerman's Encyclopedia of Healing Herbs & Spices*. Paramus, NJ: Prentice-Hall, 1996, p. 43.

BASIL

Basils, which belong to the Lamiaceae, possess the square stems, two-lipped flowers, and profusion of fragrance-bearing oil glands typifying members of the mint family. Three compounds are primarily responsible for its pungent scent: linalool, methyl chavicol, and eugenol. The relative proportion of these compounds varies between species—and hybridization introduced by individual growers—resulting in differences in scent (pronounced accents include anise, clove, cinnamon, and lemon). The more popular varieties include

- *Ocimum* 'African Blue'—reaches three feet in height, possessing purple-tinged green leaves with purple undersides and six-inch lavender flower spikes.
- *Ocimum basilicum* 'Cinnamon'—three feet tall with a purplish stem, green leaves, and lavender flower spikes. Notable for its spicy cinnamon flavor.
- *Ocimum basilicum* 'Genovese Verde Migliorato'—thirty-six inches in height, green, puckered leaves, white flowers.
- *Ocimum basilicum* 'Thai Purple'—two feet tall, possessing lavender flowers, purple-tinged stem, and long green leaves mottled with purple. Anise-flavored.

- *Ocimum basilicum* 'Dark Opal'—two feet in length with large leaves, toothed margins, and whitish-pink flowers. Besides being used for pesto as is the Genovese strain, it makes a highly prized vinegar.
- *Ocimum citriodorum* 'Lemon'—thirty inches tall with light green leaves and small white flowers.

The name *Ocimum basilicum* is derived from the Greek *okimon,* meaning smell, and *basilikos,* for king.

Native to Africa, Asia, the Caribbean, the Middle East, and South America, basil was imported into Europe approximately 2,000 years ago. In prescribing it for headaches, Dioscorides (b. 40 A.D.) warned against eating large quantities "as it dulleth the eyesight, breedeth wind, provoketh urine, drieth up milk and is difficult to digest." Pliny classified it as an aphrodisiac; it was fed to horses and donkeys during mating season. Early physicians considered it an antidote to melancholy. For centuries, basil was considered an effective remedy to inflammations of the body as well as scorpion bites. The *Virginia Gazette* carried advertisements regarding its medicinal benefits as early as 1775.

Whereas the Greeks equated basil with hatred, it symbolized love in Italy; Italian women wore a sprig of the plant to engender sympathy. It was also an established part of the courtship ritual in Romania.

The plant has long had religious as well as supernatural associations. Basil was considered a sacred herb among the Hindus and other religious sects of India. It is said to have grown around the base of Christ's tomb. Historical accounts have been split, regarding it as either sacred or evil. John Gerard, in his *Herball; or, Generall Historie of Plantes* (1597), advised readers against taking the herb internally.

Basil has enjoyed immense popularity in the present day as a flavor enhancement to meat, pasta, seafood, soups, eggs, and salads. It is perhaps most widely known as the central ingredient in pesto sauce. This delicacy originated in Italy; localities there have created their own variations in flavor.

Basil has been found to possess antimicrobial properties, leading to its use in medicinal products such as mouthwash. It is also an ingredient in certain perfumes and natural insecticides.

The culinary basils are cultivated as annuals in much of Canada and the United States; they are perennial, however, in the warmer climates where they originated. The plant requires at least several hours of full sun daily and an abundance of water. It is susceptible to fungal diseases *Fusarium oxysporum* and *Erwinia,* both of which cause sudden wilting and brown streaking on the stem (as well as leaf spotting in the latter case).

The best flavor is derived from fresh basil. Picked plants can be preserved for at least a week by placing the stems in water and refrigerating with a plastic cover. Long-term preservation can be achieved through a number of methods:

- Drying by bundling stems together with a twist tie and hanging them in a shady setting possessing good air circulation
- Chopping up leaves and freezing them in oil in ice cube trays
- Freezing leaves individually on a cookie sheet prior to packing them in a plastic container

Acknowledging its widespread popularity, McCormick's *Flavor Trends 2000* cited basil as a "Top 10 Flavor for 2000."

BIBLIOGRAPHY

DeBaggio, Thomas, and Belsinger, Susan. *Basil, an Herb Lover's Guide.* Loveland, CO: Interweave Press, 1996.

Farrell, Kenneth T. "Basil, Sweet." In: *Spices, Condiments, and Seasonings.* Westport, CT: AVI, 1985, pp. 37-41.

Herb Society of America. "Culinary Basils," www.herbsociety.org. April 1999. 4p.

McVicar, Jekka. *Herbs for the Home.* New York: Viking Studio Books, 1994.

Reader's Digest Association. *Reader's Digest Magic and Medicine of Plants.* New York: Reader's Digest Association, 1986.

BELLADONNA

Known as "deadly nightshade" due to its potentially poisonous properties if administered in too strong a dosage, belladonna *(Atropa belladonna)* is a perennial reaching as much as six feet in height. Its thick, creeping rootstock gives rise to a smooth, branched stem bearing unequal-sized pairs of dull green leaves. Roughly seven-and-one-half inches in length, the leaves bear solitary bell-shaped, purplish-brown drooping flowers from midsummer to early fall, followed by shiny black berries.

Belladonna's medical value—in this case, as a treatment for heart arrhythmia—was first noted in a book by the Greek physician Galen. Dioscorides recommended it several centuries later in his own herbal for such conditions as acid indigestion, erysipelas, heartburn, migraines, and shingles. In *Le Grand Herbier,* an herbal published in Paris in 1504, several French doctors prescribed belladonna as a potent sedative for insomnia. John Gerard's *The Herball; or, Generall Historie of Plantes* (1597) recommended it for gout and rheumatoid arthritis.

For the reduction of pain in such maladies, a fluid extract derived from belladonna rootstock and leaves is typically employed. Externally, a liniment rub composed of the herb and gin, brandy, or vodka is applied to relieve aches and pains.

Belladonna was also utilized by Samuel Hahnemann (1755-1843), the father of homeopathy, for curing scarlet fever. The treatment was continued by his disciples and was widely used by mainstream physicians up to the end of the nineteenth century.

BIBLIOGRAPHY

"Belladonna." *Medicinal Herbs Online*. Internet Enterprises, 2001. 3p. www.egregore.com/herbs.

Heinerman, John. "Belladonna *(Stropa belladonna)*." In: *Heinerman's Encyclopedia of Healing Herbs & Spices*. Paramus, NJ: Prentice-Hall, 1996, pp. 58-59.

Hutchins, Alma R. *Indian Herbalogy of North America*. Boston and London: Shambhala Publications, 1991, pp. 43, 132-135.

BILBERRY

Bilberry *(Vaccinium myrtillus)* is a deciduous shrub that can be found in cool, temperate locales throughout the Northern and Southern Hemispheres. It can be found in high elevations across Europe,

Asia, and the Rocky Mountain range in North America, thriving in damp, acid soils, woodlands, and sandy and rocky soils. It reaches roughly one foot in height and features ovate leaves that bear globular pinkish bell-like flowers in the spring. The plant's sweet, plump blue-black fruit is harvested for commercial use from July through September, most notably in Europe.

Bilberries have been valued by Europeans and Native Americans for centuries as a valuable form of nutrition. In Great Britain, they are consumed with milk and used in pies, tarts, syrups, jellies, and wine.

The earliest reference to the fruit as an herbal medicine was Hildegard of Bingen's prescription in the twelfth century for inducing menstruation. By the sixteenth century, Germans were using bilberries for bladder stones and liver disorders, and syrup derivatives for coughs and lung ailments. Two centuries later, European health care specialists were also utilizing the fruit to treat typhoid fever, gout, rheumatism, and infections of the mouth, skin, and urinary tract. By the twentieth century, it had become popular as an astringent and disinfectant mouthwash for mouth inflammations, and a tea made from the dried berries was employed as a tonic and to curb diarrhea and bleeding, promote urination, and prevent scurvy.

A revival of interest in the medicinal properties of bilberry occurred during World War II when British Royal Air Force pilots noted improved night vision during bombing missions after eating bilberry jam. Laboratory tests in the 1960s and later clinical studies led to the following findings relating to the use of bilberry fruit extracts:

- Relief of disorders of the peripheral blood vessels, particularly capillary fragility associated with aging
- Reduction of bruising, blood in the stool, and swelling due to water retention
- Relief from feelings of heaviness, pain in the legs and ankles, and sensations of burning, prickling, or numbness of the skin in subjects suffering from water retention in lower limbs or varicose veins
- Protection of capillaries from free-radical damage, stimulation of healthy connective tissue formation, and promotion of new capillary formation
- Reduction of blood platelet stickiness, which sometimes leads to heart attacks or stroke

- Increased adaptation to darkness due to an enhanced regeneration rate of the purple pigment rhodopsin in the retina
- Deterrence of enzyme reactions that damage the retina
- Reduction (or even disappearance) of hemorrhages in the retina

Furthermore, studies have shown that bilberry tea aids in relieving diarrhea and inflammation of the mucous membranes of the mouth and throat.

The dated results of these findings, combined with the fact that most studies were conducted by French or Italian researchers and published in foreign scientific journals, has limited their acceptance in English-speaking nations. On the other hand, bilberry's safety is widely accepted with no reported contraindications, interactions with other drugs, or side effects reported in reputable studies. Therefore, bilberry dietary supplements—e.g., extracts standardized to 25 percent anthocyanosides and tablets and capsules of the dried fruits—are readily available in the United States. Health food outlets also carry bilberry tea, produced from twenty to sixty grams of dried ripe berries with a recommended three dosages daily. Evidence of the increasing popularity of bilberry is its inclusion in Discount Natural Herbs' "Top Ten Selling Herbs" through mid-2001.

BIBLIOGRAPHY

Cunio, L. *"Vaccinium mytillus." Australian Journal of Medical Herbalism* 5(4) (1993): 81-85.

Foster, Steven. "Bilberry: Food and Medicine." *The Herb Companion.* (December/January 1996/1997): 68-69.

Garland, Sarah. *The Complete Book of Herbs & Spices.* Pleasantville, NY: Reader's Digest Association, 1993.

Morazzoni, P., and Bombardelli, E. *"Vaccinium mytillus." Fitoterrapin 67*(1) (1996): 3-29.

BLACK COHOSH

Black cohosh *(Cimicifuga racemosa)* is a wildflower native to the moist or dry woods of eastern North America. Cohosh is an Algonquian word meaning "rough"; it came from the plant's lumpy blackish rhizomes. Another popular name, rattletop, was a result of the sound the dry seeds make in their pods atop the flower stalks. The species name is derived from the Latin *cimex,* a kind of bug, and *fugare,* "to put to flight." It is believed that the plant's strong odor repels insects. *Racemosa,* literally "in the form of a raceme," refers to the arrangement of individual flowers on an elongated stalk.

The mature black cohosh features clumps of resilient stems three to eight feet in height. Large, alternate green leaves are pinnately compound with toothed leaflets. During the summer months, long wands of small, starry white flowers extend beyond the foliage. The flowers, thought to be pollinated by green flesh flies, possess no petals; the greenish white sepals fall off soon after a flower opens, leaving a tuft of showy stamens surrounding a single pistil.

The rhizome has been used by Native Americans to relieve menstrual cramps as well as ease childbirth. The plant was an ingredient of Lydia Pinkham's Vegetable Compound, a patent remedy for "female complaints" popular in the nineteenth and early twentieth centuries. Modern research has revealed that the herb contains plant estrogens, rendering it a natural alternative to hormone replacement therapy. Eight German clinical studies—supported in recent years by the observations of American doctors—have indicated that black cohosh is safe for use during menopause without significant side effects.

Long considered both a relaxant and mild tonic, black cohosh has also been used to treat arthritis, coughs, diabetes, tinnitus, dropsy, neuralgia, malaria, yellow fever, and lung disorders. The rhizomes have been employed in poultices to treat snakebite, hence it is commonly referred to as black snakeroot.

Research studies have documented that extracts of the rhizome are anti-inflammatory and capable of lowering blood pressure in laboratory animals. The U.S. Food and Drug Administration, however, has reported finding "no pharmacological evidence of the therapeutic value" (*Herb Quarterly,* 1999, p. 8). Nevertheless, John Heinerman believes that the herb's track record for reducing hypertension may make it useful in treating women beset by Premenstrual Tension Syndrome.

The flowers and seed heads of black cohosh (and related species) are widely utilized as cut flowers. It is also popular as the backdrop to a shady border or woodland flower garden, blending well with ferns and various coarse-leaved plants in particular.

BIBLIOGRAPHY

"Black Cohosh: Safe for Menopause." *The Herb Quarterly* (Winter 1999): 7-8.

Heinerman, John. "Black Cohosh *(Cimicifuga racemosa).*" In: *Heinerman' s Encyclopedia of Healing Herbs & Spices.* Paramus, NJ: Prentice-Hall, 1996, pp. 72-73.

Strauch, Betsy. "An Herb To Know." *The Herb Companion* (October/November 1995): 24-25.

BLACK PEPPER

Black pepper *(Piper nigrum)* is presently consumed more than all other spices combined. For several centuries, it was the primary staple of world trade. Throughout much of the Middle Ages, peppercorns were so costly that they served as hard currency. Taxes, dowries, and other tributes allegedly included levies of pepper. Dockmen unloading pepper ships were not allowed pockets in their clothing due to concerns that they might attempt to steal some corns.

Native to India, black pepper today is cultivated extensively in tropical climates worldwide, most notably in Indonesia. Black, green, and white pepper represent three stages of ripeness in the pea-sized berries produced by the climbing bush.

In addition to its universal appeal as a preservative and seasoning for food, black pepper possesses a number of medicinal uses. It has long been considered an aid to digestion; as early as 400 B.C., Hippocrates declared that it assisted the functioning of gastric juices. In Egypt, a decoction of the spice has long been used as a mouthwash to ease toothache pain. A Philippine folk remedy consists of warming equal amounts of black pepper and anise with brandy to reduce high body temperature due to fever.

John Heinerman relates that a Miami policeman, responsible for disposing of voodoo sacrifices left at the Dade County Courthouse by relatives of defendants on trial for alleged crimes, indicated that pepper is one of various foods associated with voodoo rituals. For instance, black pepper is employed to keep someone in jail for a lengthy period of time.

BIBLIOGRAPHY

Charalambous, George (Ed.). *Spices, Herbs and Edible Fungi.* Amsterdam: Elsevier, 1994.

Farrell, Kenneth T. "Pepper, Black." In: *Spices, Condiments, and Seasonings.* Westport, CT: AVI, 1985, pp. 178-183.

Heinerman, John. "Black Pepper *(Piper nigrum).*" In: *Heinerman's Encyclopedia of Healing Herbs & Spices.* Paramus, NJ: Prentice-Hall, 1996, pp. 74-75.

BOUQUET GARNI

Bouquet garni is a French term for "an herb bundle" used in cooking. Its contents generally include

- Five sprigs of parsley
- Two sprigs of thyme
- One bay leaf
- A portion of dried orange peel
- One sprig of sweet marjoram
- One sprig of basil
- Two celery leaves
- A small piece of cinnamon stick
- A clove of garlic
- A small blade of mace
- A pod of red pepper

The ingredients can vary, however, according to individual tastes. In order to facilitate removal after cooking, the herbs are either tied together or placed in a muslin sack.

The bouquet garni is used primarily as a flavor enhancer. In addition, it acts as an aid to digestion. Its presence lessens the likelihood of constipation, heartburn, and intestinal gas. These benefits are not realized with commercially produced "mixed herb" preparations or powdered bouquet garni available in packets.

BIBLIOGRAPHY

Farrell, Kenneth T. *Spices, Condiments, and Seasonings*. Westport, CT: AVI, 1985.

Heinerman, John. "Bouquet Garni." In: *Heinerman's Encyclopedia of Healing Herbs & Spices*. Paramus, NJ: Prentice-Hall, 1996, p. 90.

BROOKLIME

Brooklime *(Veronica beccabunga)* possesses thick, succulent stems that grow azure-blue flowers from May through late August. The leaves are opposite, short-petioled, oblong to ovate, and either crenate or serrate. Its taste is reminiscent of watercress, albeit more bitter, hence the nickname "mouth smart." It generally takes root in shallow streams, babbling brooks, and similar waterways.

Old British herbals cite brooklime for its effectiveness in treating hemorrhoids, diabetic leg ulcers, and skin sores. The expressed juice from the crushed herb may be taken orally or applied directly to the problem itself; mixing the herb with water in a food processor is an efficient means of obtaining a sufficient amount of the juice.

Mary Thorne Quelch, in *Herbs for Daily Use,* touted its value in treating skin, eye, and head conditions. She stated,

> Our forefathers had great faith in a medicine called Spring Juice, which, taken freely in the early part of the year, was held to be without rival in clearing the blood and improving the digestion . . . A wineglassful of this mixture taken fasting each morning for a week was guaranteed to clear the complexion and brighten the eyes, besides curing persistent headaches and heal-

ing skin troubles. There is every probability that the draught deserves its high reputation. (Cited in Heinerman, 1996, p. 96-97)

Also known as water pimpernel, brooklime was an extremely popular blood-purifying herb in Europe and Asia up until the mid-twentieth century. Although still employed to some extent due to its wide-ranging abundance, it is no longer in favor with practicing herbalists.

BIBLIOGRAPHY

"Brooklime." *Medicinal Herbs Online.* Internet Enterprises, 2001. 3p. www.egregore.com/herbs.

Grieve, M. "Brooklime." *A Modern Herbal.* Electric Newt, 1995. 2p. www.botanical.com.

Heinerman, John. "Brooklime *(Veronica beccabunga).*" In: *Heinerman's Encyclopedia of Healing Herbs & Spices.* Paramus, NJ: Prentice-Hall, 1996, pp. 96-97.

Quelch, Mary Thorne. *Herbs for Daily Use.* London: Faber and Faber, Ltd., 1941.

BURDOCK

There are two main classes of burdock. Common burdock *(Arctium lappa)* is commonly found on dry roadsides and wastelands in temperate zones. A biennial reaching more than three feet in height, it possesses big, oval leaves that are dull green on top and downy underneath. Small, purple flower heads appear in mid-summer; its sharply hooked bracts evolve into clinging burs.

The leaf stems are peeled and added to soups. The dried leaves can be mixed with dandelion to make beer. The herb also has a variety of medicinal uses. A leaf infusion or decoction of the root can cleanse the blood and, as a hair rinse, controls dandruff. Boiled or crushed leaves are effective in the treatment of aches and bruises.

Greater burdock *(Arctium lappa)* is often called cockleburr due to its round, brown bristly burrs. It is widely cultivated in Japan as a food source in much the same manner that we use carrots.

Greater burdock root is generally felt to be the most widely used herbal blood purifier. John Heinerman considers it "*the* most important herb for treating chronic skin problems," (p. 103) including acne, boils, cankers, carbuncles, eczema, herpes, psoriasis, styes, syphilitic sores, and similar conditions. A burdock-catnip tea has been found effective in getting rid of persistent kidney stones and gallstones. Alternative-care physicians regularly prescribe dishes containing burdock root to patients recuperating from illness or surgery, or simply to ensure greater vigor and vitality.

BIBLIOGRAPHY

Grieve, M. "Burdock." *A Modern Herbal*. Electric Newt, 1995. 4p. www.botanical.com.

Heinerman, John. "Burdock *(Arctium lappa)*." In: *Heinerman's Encyclopedia of Healing Herbs & Spices*. Paramus, NJ: Prentice-Hall, 1996, pp. 103-106.

C

CALENDULA

Considered a weed by many because it spreads easily in gardens and waste areas, calendula *(Calendula officinalis)* yields distinctive yellow to orange flower heads. According to an old folk belief, if its flower heads close up after seven in the morning, rain is guaranteed the next day.

The plant is much more popular in Europe than in North America, particularly in a salve or dilute tincture for treating virtually any kind of external skin, muscle, or blood problems—for example, wounds, sores, burns, frostbites, varicose veins, pulled muscles, boils, bruises, sprains, athlete's foot, etc. In the United States, Calendula Dairy Salve, marketed by Old Amish Herbs, with a base of pure pork lard (one of the most absorbent materials for the skin), is a widely used product adapted from time-tested folk remedies.

The Soviet journal *Vatreshni Bolesti* (Volume 20, June 1981, Heinerman, 1996, p. 112) published the results of two intensive medical studies that confirmed the value of calendula in healing duodenal ulcers, inflammation of both the stomach and duodenum, and intestinal colitis. Such findings appear to demonstrate the clinical validity of the herb in treating all types of inflammation.

Anecdotal evidence regarding calendula's value also abounds. A German herbalist, Maria Treben, relates that a homemade healing salve developed from fresh calendula leaves, flowers, and stems turned legs covered with varicose veins to clear, smooth skin in four weeks.

The herb is also prized in Europe as a flavoring ingredient. John Heinerman offers an instance where calendula flowers imbue butter beans with a more delicate, creamy flavor, especially when complemented by a rich sauce. General public appreciation of calendula is reflected by its inclusion in Deborah C. Harding's *The Green Guide to Herb Gardening: Featuring the 10 Most Popular Herbs* (2000).

BIBLIOGRAPHY

Gardner, Jo Ann. "Calendula (Pot-Marigold)." In: *Living with Herbs*. Woodstock, VT: Countryman Press, 1997, pp. 124-127.

Harding, Deborah C. *The Green Guide to Herb Gardening: Featuring the 10 Most Popular Herbs*. St. Paul, MN: Llewellyn, 2000.

Heinerman, John. "Calendula *(Calendula officinalis)*." In: *Heinerman's Encyclopedia of Healing Herbs & Spices*. Paramus, NJ: Prentice-Hall, 1996, pp. 111-113.

CAYENNE PEPPER

A perennial in the American tropics where it is native, cayenne pepper *(Capsicum frutescens; Capsicum annuum)* is an annual when cultivated elsewhere. It is one of many capsicum varieties (often used interchangeably), including paprika, red pepper, Tabasco, chili, pimento, and sweet pepper. Reaching three or more feet in height, its glabrous stem is woody at the bottom and branched near the top and sustains ovate to lanceolate leaves. The drooping, white to yellow flowers—which grow alone or in pairs of three during April to September—give rise to a many-seeded, yellow to red pod (the ripe fruit, or pepper).

Although historically associated with the exploration of the New World, cayenne pepper has been consumed by humans at least as far back as 7500 B.C. It was first discovered by Europeans in what is now French Guiana on one of several voyages to America by Columbus. West Indian natives used it primarily to season meats. During early colonial times in Jamaica, red pepper juice was placed in the eyes of slaves to punish misdemeanors. It was mixed with honey and lentils to remove facial spots. It was also utilized as a stimulant for sluggish digestive systems and to treat sore throats, malaria, colic, and alcoholism.

Capsaicin has been identified as the compound that endows chili peppers with their distinctive fiery properties. The widely used Official Chili Pepper Heat Scale, which utilizes a 0-10 rating scale, gives Cayennes (along with Tabascos) an eight, hotter than Jalapenos (five) but behind the Scotch Bonnet and Habanero (both ten).

Capsaicin has been shown to lower the body's temperature. This is accomplished by stimulating, then desensitizing, the warmth detectors in the hypothalamus gland. In order to tolerate heat, natives in hot climates in Latin America and Africa consume chili peppers in large quantities.

European studies in the 1980s concluded that consuming cayenne pepper boosts vitamin C levels as well as speeding up the body's metabolism. It appears that capsaicin "resets" individual "fat thermostats," causing the body to burn more fat through chemical combustion instead of storing it in muscle tissue. Members of the chili pepper family contain much more vitamin C per unit weight than citrus fruit. Because they prevented scurvy (a vitamin C deficiency disease once prevalent on ships), Spanish sailors brought along pickled peppers during voyages from the sixteenth century onward. Today, Super C, produced by Old Amish Herbs, utilizes cayenne to enhance the effectiveness of vitamin C.

Jewish grandmothers living in Brooklyn, New York, have been known to push hot chicken soup laced with cayenne pepper and chopped garlic clove as the most effective means of fighting cold and flu symptoms. Often referred to as "Jewish penicillin," it is frequently advocated over antibiotics by physicians.

Cayenne pepper's massive popularity as a flavoring agent is reflected by its inclusion in McCormick's *Flavor Trends 2000* "Top 10 Flavors for 2000." The spice provides many additional medicinal benefits as well. Most notably, it stops bleeding quickly; relieves rheumatoid arthritis pain (through application of topical creams containing capsaicin); reduces the risk of blood clots and strokes; heals stomach ulcers; brings down blood sugar levels; lowers cholesterol and the associated risk of heart disease; stops cluster headaches; helps keeps extremities warm in wintertime; increases the body's energy levels; treats sprains and bruises (by means of an ointment developed in China that combines ground hot pepper and Vaseline); cures sore throat; decreases pain associated with shingles, trigeminal neuralgia, and diabetic neuropathy; and reduces the risk of contracting pneumonia and other respiratory system diseases.

BIBLIOGRAPHY

Bosland, Paul W. "Chiles: History, Cultivation, and Uses." In Charalambous, George (Ed.), *Spices, Herbs, and Edible Fungi.* Amsterdam: Elsevier, 1994, pp. 347-366.

Farrell, Kenneth T. "Capsicum (Cayenne or Red Pepper)." In: *Spices, Condiments. and Seasonings.* Westport, CT: AVI, 1985, pp. 50-54.

Heinerman, John. "Cayenne Pepper *(Capsicum frutescens, Capsicum annuum).*" In: *Heinerman's Encyclopedia of Healing Herbs & Spices.* Paramus, NJ: Prentice-Hall, 1996, pp. 127-134.

CHAMOMILE

The popularity of chamomile *(Matricaria chamomilla)* reaches back to antiquity. The Egyptians called it the "Flower of the Sun God" due to its diaphoretic features, shown to help curb fever. The Greeks also valued the plant—known as "ground apple" for its scent—as a medicinal. In the Middle Ages, the Anglo-Saxons considered it one of nine sacred herbs provided to man by the god Woden.

The flower—which can be harvested during the summer months—constitutes the useful portion of the plant. Its active constituents include volatile oil (a dark blue color when steamed distilled), flavonoids, coumarins, plant acids, fatty acids, cyanogenic glycosides, salicylates, choline, and tannins. Similar in its effects to Roman chamomile *(Chamaemelum nobile),* the herb serves as an analgesic, anti-inflammatory, antihistamine, antiseptic, antispasmodic, bitter tonic, decongestant, diaphoretic, diuretic, and sedative.

Chamomile has proven effective in a wide range of applications. Ingested as a tea, it can treat menstrual and other muscle cramps, upset stomach, excess gas, diarrhea, constipation, fever, congestion, hemorrhoids, nervous tension, poor appetite, digestive problems, headache, insomnia, ulcers, nausea and vomiting, dysmenorrhea, and edema. In a sitz bath, the tea sooths cystitis and external hemorrhoids. Used as a poultice, it can heal burns, infections, rashes, and skin ulcers; it also functions as a hair tonic. Steam inhalation eases asthma, hay fever, sinusitis, and catarrh, and the essential oil can be applied externally for eczema and neuralgia.

Its versatility has rendered it one of the most widely used herbs today. This is reflected by its inclusion in Deborah C. Harding's *The Green Guide to Herb Gardening: Featuring the 10 Most Popular Herbs* (2000).

BIBLIOGRAPHY

Foster, Steven. "Indigestion? Spell Relief the Herbal Way." *The Herb Companion* (October/November 1996): 55-57.

Gardner, Jo Ann. "Chamomile." In: *Living with Herbs*. Woodstock, VT: Countryman Press, 1997, pp. 129-131.

Harding, Deborah C. *The Green Guide to Herb Gardening: Featuring the 10 Most Popular Herbs*. St. Paul, MN: Llewellyn, 2000.

Hollis, Sarah. "Chamomile." In: *The Country Diary Herbal*. Rochester, VT: Healing Arts, 1990, pp. 38-40.

Langer, Robin E. "Herb-of-the-Month: Chamomile *(Matricaria Chamomilla)*." *Snowbound Herbals* (January 2001): 2p. www.sbherbals.com.

CHAPARRAL

In his book, *Desert Happy,* naturalist Douglas Rigby terms chaparral or, as it is sometimes called, the creosote bush, "a plant genius." It flourishes in the nutritionally deficient soil of the American southwest, retaining its characteristic bronze to mustard-green hue in the face of oppressive heat and as little as a few tablespoons of water per year. In addition, it secretes a powerful antigrowth substance that discourages invasive vegetation from taking root nearby.

John Heinerman posits that a group of creosote bushes in the Mojave Desert, 150 miles northeast of Los Angeles, are the oldest living organisms on the planet—11,700 years old! The plant's durability is a result of its nordihydroguaiaretic acid (NDGA) content. Used in numerous food products to prevent rancidity, NDGA is a powerful antioxidant capable of holding in check the free radicals (molecular fragments within the human body responsible for causing blood clots, arthritis, senility, and greatly hastening the aging process). Heinerman argues that "while chaparral may not hold quite the same promises expected of ginseng for longevity, it can certainly help to slow down the aging process quite a bit from the foods we eat on a daily basis" (p. 143).

Both Mexican-Americans and the Pima Indians of the southwest boil chaparral in alcohol (whiskey or wine) for use as a dandruff remover that far outperforms national brand shampoos. This solution can be applied to cats and dogs bothered by lice or fleas. It can also be employed as a douche to fight yeast infection, and soaking one's feet in it has been found effective in curing athlete's foot.

Dr. Charles R. Smart, Chief of Surgery at Salt Lake City's LDS Hospital until 1985 and an internationally known cancer specialist, noted in the June 1978 issue of *Herbalist* that "chaparral tea produced regression of tumors but not necessarily cures" (Heinerman, 1996, p. 143). One of his patients, an eighty-five-year-old named Ernest Farr, was afflicted with advanced melanomas on his face and neck. His cancer virtually disappeared after a treatment regimen that included no medication other than chaparral tea; he died of other causes at the age of ninety-six. Based on the success of this herb solution in treating his wife's tumor, Tom Murdock formed what would become the largest health food concern in the nation, Nature's Way. The company's first product was chaparral capsules. A number of publications, including the American Cancer Society's *Unproven Methods of Cancer Management* (1970), have further documented chaparral's ability to hold various cancers in check, most notably metastatic choriocarcinoma, lymphosarcoma, and leukemia.

BIBLIOGRAPHY

American Cancer Society. *Unproven Methods of Cancer Management.* American Cancer Society, 1970.

Heinerman, John. "Chaparral *(Larrea divericata).*" In: *Heinerman's Encyclopedia of Healing Herbs & Spices*. Paramus, NJ: Prentice-Hall, 1996, pp. 140-145.

Rigby, Douglas. *Desert Happy*. New York: Lippincott, 1957, pp. 108-109.

CHERVIL

Chervil *(Anthriscus cerefolium)* is an annual possessing a finely grooved, branched stem growing twelve to sixteen inches tall from a thin, whitish root. The leaves are light green, opposite, and bipinnate; the lower ones petioled, the upper sessile on stem sheaths. Small, white flowers—appearing as compound umbels from May to July—are followed by elongated, segmented seeds during August and September.

Chervil is a variant of the Greek term denoting "leaf of rejoicing" or "cheer-leaf." Of East European origin, it now thrives in the wild areas of Iran and southern Russia. The ancient Romans imported the plant as far west as Great Britain. Long a symbol of sincerity, legend holds that chervil sharpens a dull wit, prods the memory, and restores some measure of youth to the aged. English herbalist Gerard would write in the sixteenth century: "It is good for old people—it rejoiceth and comforteth the heart and increaseth their strength." Although cultivated in the United States and much of Europe, it is now used primarily in France for seasonings, salads, soups, and as a pot-herb.

According to French folk healer Maurice Messegue, ancient cultures used chervil for a variety of eye disorders. Much of Europe, particularly France, has continued to employ it with great success for ophthalmitis (severe inflammation of the deeper structures of the eye), detached retina, cataracts (loss of eye lens transparency), conjunctivitis, and sometimes even glaucoma. Respected practitioners such as Leon Binet, former Dean of the Paris Faculty of Medicine and author of many medical books, have actively advocated its use. John

Heinerman notes that when chervil is combined with other herb remedies such as eyebright, "the results are nothing short of simply amazing" (p. 146).

BIBLIOGRAPHY

Farrell, Kenneth T. "Chervil." In: *Spices, Condiments, and Seasonings*. Westport, CT: AVI, 1985, pp. 73-75.

Gardner, Jo Ann. *Living with Herbs*. Woodstock, VT: Countryman Press, 1997.

Heinerman, John. "Chervil *(Anthriscus cerefolium)*." In: *Heinerman's Encyclopedia of Healing Herbs and Spices*. Paramus, NJ: Prentice-Hall, 1996, pp. 145-147.

CHICKWEED

Chickweed *(Stellaria media)* is an extremely hardy annual that flowers throughout the country during the entire year. Generally appearing in the fall, it is capable of thriving through winter storms in the far north and cannot be eradicated by most weed killers.

Reaching a foot in height with matted to upright stems, chickweed possesses egg-shaped lower and median leaves and stemless and highly variable upper leaves. The bright white flowers open under the sun, while typically gathering themselves together at night as well as on cloudy or foggy days.

John Heinerman states that chickweed "ranks beside herbs such as burdock root as being terrific blood cleaners" (p. 147). Chickweed capsules or a tea extract are effective in treating blood poisoning or tetanus caused by chemical dye or dirt within the bloodstream, carbuncles, boils, venereal disease, herpes sores, and swollen testicles and breasts. A salve made from freshly cut or powdered chickweed also provides relief for chronic itching and severe rashes.

Mike Tierra, a licensed herbalist based in Santa Cruz, California, notes in his book, *Way of Herbs,* that "chickweed is particularly useful for reducing excess fat, having both mild diuretic and laxative properties" (Heinerman, 1996, p. 148). In a weight-reducing program, the plant can be ingested both in greens and as an herbal tea.

BIBLIOGRAPHY

Garland, Sarah. *The Complete Book of Herbs & Spices*. Pleasantville, NY: Reader's Digest Association, 1993.

Heinerman, John. "Chickweed *(Stellaria media)*." In: *Heinerman's Encyclopedia of Healing Herbs & Spices*. Paramus, NJ: Prentice-Hall, 1996, pp. 147-149.

Tierra, Mike. *Way of Herbs*. Edited and supplemented by Subhuti Dharmananda. Santa Cruz, CA: Unity Press, 1980.

CINNAMON

One of the best known spices for thousands of years, cinnamon is derived from the bark of the cinnamon tree *(Cinnamomum zeylanicum)*. The most highly prized bark is harvested from young shoots, the paler the better. Cinnamon is often confused with cassia (or Chinese cinnamon); the latter is native to Burma and possesses a more pungent taste best suited to spiced meat, pilaus (rice or cracked wheat with boiled meat and spices), and curries. On the other hand, true cinnamon comes from Sri Lanka (formerly Ceylon) and has a more delicate flavor ideal for breads, breakfast cereals, cakes, cookies, pastries, and other sweet dishes.

Cinnamon was evidently a fixture at least as far back as the days depicted in the Old Testament. It was a major ingredient in a "holy annointing oil" used by Moses. Due to its popularity, the cinnamon tree is cultivated in warmer climates worldwide, most notably in the West Indies and East Asia. The demand for cinnamon is such that McCormick's *Flavor Trends 2000* included the spice in its "Top 10 Flavors for 2000."

Besides being held in high esteem as a flavoring, cinnamon has many other contemporary uses. Among other things, it can serve as

- An antiseptic mouthwash
- A remedy for acid indigestion, heartburn, and cramps

- A cold and flu fighter
- A remedy for yeast and fungal (e.g., athlete's foot) infections
- An agent for the prevention of cancer

While some of these applications grew out of old folk medicine practices, recent studies have revealed a scientific basis for the medicinal use of cinnamon. Two cancer specialists based at the British Columbia Cancer Research Centre in Vancouver have identified cinnamic acid as the active ingredient in the spice which helps prevent cancer induced by chemicals present in much of today's food products. In 1974, the *Journal of Food Science* (cited in Heinerman, 1996, p. 161) reported that as little as 2 percent or twenty milligrams of the spice per milliliter of a yeast-extract and sucrose broth inhibited 97 to 99 percent of existing toxic molds.

BIBLIOGRAPHY

Farrell, Kenneth T. "Cinnamon, Ceylon." In: *Spices, Condiments, and Seasonings*. Westport, CT: AVI, 1985, pp. 80-85.

Garland, Sarah. *The Complete Book of Herbs & Spices*. Pleasantville, NY: Reader's Digest Association, 1993.

Heinerman, John. "Cinnamon *(Cinnamomum zeylanicum)*." In: *Heinerman's Encyclopedia of Healing Herbs & Spices*. Paramus, NJ: Prentice-Hall, 1996, pp. 160-162.

Moyler, D.A. "Cinnamon (bark)." In: Charalambous, George (Ed.), *Spices, Herbs and Edible Fungi*. Amsterdam: Elsevier, 1994, p. 60.

Vernin, C., Metzger, J., Pujol, L., and Parkany, C. "GC/MS Analysis of Cinnamon and Cassia Essential Oils: A Comparative Study." In: Charalambous, George (Ed.), *Spices, Herbs and Edible Fungi*. Amsterdam: Elsevier, 1994, pp. 411-425.

CLOVES

One of the most popular spices throughout history, cloves *(Syzygium aromaticum)* are a product of an evergreen tree native to the Moluccas islands. The clove tree thrives only in a tropical climate near the sea, reaching a height of thirty feet. The leaves resemble bay and the flower buds are pinkish-red in color; when open, the latter reveal yellow petals and a profusion of stamens. The buds—harvested and dried in the sun until they turn brown—are the useful portion of the plant.

Cloves were listed in early Chinese herbals and widely used during classical antiquity. Long before Christ's birth, the upper classes sucked on them to perfume their breath. Early physicians prescribed the spice as an aid to digestion, believing that it strengthened the stomach, liver, and heart. The widespread use of cloves in the English diet was ushered in by the Norman invasion of 1066; however, the spice remained expensive until the Portuguese broke the Venetian spice monopoly several hundred years later. The Dutch seized the Moluccas in the mid-seventeenth century, thereby suppressing the export of clove seed or seedings until the next century.

The island of Zanzibar is appreciated for its many fragrant clove trees. Tom Stobart, in his *Herbs, Spices and Flavorings,* notes, "On a hot, muggy evening when the light breezes filter through the trees, if one approaches the island from downwind, one can smell cloves even before the land comes in sight" (cited in Heinerman, 1996, p. 168).

As a result of their fragrance and sharp taste, cloves are widely used to season meat, vegetable dishes, soups, and some desserts such as cake and cookies. The flavor of red wine can be intensified by heating and adding several buds. A spicy tea that refreshes and stimulates the appetite is produced by crushing several buds and adding them to boiling water.

Cloves possess a number of other positive properties. Antiseptic with a mild anesthetic action, its oil can be rubbed on sore gums or applied (via a cotton wad) to an aching tooth. It can also be used as a mouth rinse for bad breath. Cloves have been a staple in air fresheners at least as far back as the sixteenth century. Sucking on one or two whole cloves has been recommended for curbing cravings for alcohol.

BIBLIOGRAPHY

Farrell, Kenneth T. "Cloves." In: *Spices, Condiments, and Seasonings.* Westport, CT: AVI, 1985, pp. 86-91.

Garland, Sarah. *The Complete Book of Herbs & Spices.* Pleasantville, NY: Reader's Digest Association, 1993.

Heinerman, John. "Cloves." In: *Heinerman's Encyclopedia of Healing Herbs & Spices*. Paramus, NJ: Prentice-Hall, 1996, pp. 168-169.

Stobart, Tom. *Herbs, Spices, and Flavorings.* Woodstock, NY: Overlook Press, 1982.

COFFEE; COFFEE SUBSTITUTES

The seed of *Coffea arabica,* a shrub of the madder family native to Ethiopia, first caught the attention of outsiders when Arab traders visiting the East African coast observed the locals chewing on a substance based upon a peanut-sized bean. Arab sailors found the herb ideal for keeping awake while on watch; they named it *qahwah,* which meant "keeps awake."

From Yemen, the "bean broth" quickly spread across Arabia in the eleventh century. With alcohol intake forbidden, coffee filled the void in Muslim religious ritual and everyday life. The drink continued to gain converts as Islam's sphere of influence spread north and west. Recognizing its potential, Venetian merchants carried coffee from Constantinople to Italy in 1615; by 1750 it had conquered much of Europe.

As the craze for the bean was spreading on that continent, the Catholic hierarchy denounced it as the brew of infidels. Pope Clement allegedly tried a cup and was converted on the spot; he resolved the matter by baptizing the drink, thereby conferring Christian status. Many European authors and artists of the eighteenth century found coffee to be a creative stimulant. Voltaire is said to have consumed fifty cups per day; the French statesman Talleyrand was moved to share his prescription for the ideal serving: "black as the devil, hot as hell, pure as an angel, sweet as love." Johann Sebastian Bach composed a cantata in tribute to the brew.

Coffee cultivation was established in the tropical Americas in the early 1700s. This region now enjoys a near monopoly on arabica beans (long the preferred type for their superior richness of flavor) thanks to favorable factors such as rich soil, reliable rainfall, and altitudes ranging from 3,000 to 6,000 feet above sea level.

A plethora of studies have appeared in recent years citing healthful benefits of coffee consumption. An American study profiled in a 1999 issue of *JAMA* found that men drinking at least two cups of caffeinated coffee daily had a 40 percent lower risk of developing gallstone disease (increased amounts of coffee intake decreased the risk to an even greater extent). Coffee has also been touted for the prevention of colon cancer, ridding the body of toxins, and fighting asthma and bronchial congestion. The grounds are even used as a body rub (particularly in Japan, where the practice has reached faddish proportions).

While coffee is now popular worldwide in large part both for its distinctive taste and its ability to stimulate the nervous system, many users have expressed concern over a host of less desirable side effects caused by the principal active ingredient, xanthine alkaloid. These effects—which include nervousness, irritability, anxiety, insomnia, disturbances in heart rate, and long-term changes in blood pressure, coronary circulation, and the secretion of gastric acids—combined with coffee's high price, have motivated consumers to find healthier, less expensive substitutes. These alternatives are generally either blended to achieve a drink that approximates coffee (sans caffeine), or mixed with coffee in order to reduce caffeine content.

Coffee substitutes are derived from the roots or seeds of other herbs. Root preparation involves washing, slicing, and drying prior to slow-roasting; when crisp and dark brown, the root is ground and infused in much the same manner as real coffee. Some of the more popular alternatives are discussed below.

Chicory *(Cichorium intybus)* was brought to the New World from Europe for medicinal and culinary uses. The plant is easily identified by its sky-blue blossoms which appear in midsummer above a rosette of dandelion-like leaves. The roots must be dug up prior to or after flowering in order to minimize bitterness. Considered a common weed on North American roadsides, the toxins derived from automobile exhausts and other wastes render it necessary to obtain the chicory used in commercial coffee blends from cultivated European varieties.

The **dandelion** (*Taraxacum* spp.) has long been popular with the rural poor who ground-roasted its taproots to extend their coffee stock. Many coffee connoisseurs believe that these roots give *Coffea arabica* a run for its money in the taste sweepstakes. Widely available, its main drawbacks are the limited time frame for gathering the roots (those obtained before the plant flowers are mildest in flavor), and the tendency of the tubers to dry to about a quarter of their original size.

The strangely named **Jerusalem artichoke** *(Helianthus tuberosus),* a perennial member of the sunflower family that produces edible tubers, was once a food staple. The plant remains prolific on roadsides and in fields and waste areas; white- and red-tuber varieties are widely cultivated in Europe as well. Possessing a delicious, nut-like taste whether served raw or cooked, the root—when ground and sweetened—is highly valued as a blending ingredient in coffee.

Some roots employed in herbal coffee blends are better known for other uses. These include beets (a source of dark coloring and carmelized sugar), carrots (sweetness), parsnips (sweetness), burdock (coloring), and salsify (bitterness).

A number of seeds are also being marketed today as coffee substitutes. These include carob, coffeeberry, some members of the pea family, bitterroot, cacao, juniper berries, and beechnuts.

BIBLIOGRAPHY

Henderson, Robert K. "Jive Java." *The Herb Companion* (February/March 2000) 3p. Also available at www.discoverherbs.com; includes a link to a two-page article titled "Seed Substitutes."

Leigh, Evelyn. "Coffee Lowers Risk of Gallstone Disease for Men in Major Study." *Herb World News Online*. Boulder, CO: Herb Research Foundation, 2000. 2p. www.herbs.org/current.

Leitzmann, Michael F., Willett, Walter C., Rimm, Eric B., Stampfer, Meir J., Spiegelman, Donna, Colditz, Graham A., Giovannucci, Edward. "A Prospective Study of Coffee Consumption and the Risk of Symptomatic Gallstone Disease in Men." *JAMA 281*(22) (1999): 2106-2112.

Starbird, Ethel A. "The Bonanza Bean: Coffee." *National Geographic 159*(3) (March 1981): 388-405.

COMFREY

While echinacea is now the first choice for fighting colds and flu among believers in alternative medicine, Americans in the late eighteenth and nineteenth centuries preferred Indian Sage, now known as comfrey *(Symphytum officinale)* or boneset. Other names over the years have included thoroughwort, feverwort, and agueweed. Medical practitioners, herbalists, and homesteaders have left considerable evidence regarding the herb's popularity. Charles Millspaugh, in his reference book *American Medicinal Plants,* stated that by the late nineteenth century, the herb hung from the rafters of nearly every woodshed and attic across the nation.

Comfrey was also employed against malaria, yellow fever, and other serious viral contagions during early American history. Dr. Benjamin Smith Barton wrote in the early 1800s that the plant played a key role in halting a 1793 yellow fever epidemic in Philadelphia and was used by a Woodbury, New Jersey, physician to treat typhus and other fevers. C.S. Rafinesque, author of an important work on medical botany published in the early 1800s, noted that the herb cured a form of influenza referred to as breakbone fever (since identified by medical historians as a variation of the mosquito-transmitted virus dengue). This illness resulted in joint and muscle pains which made patients feel as though their bones were breaking; use of the herb induced a feeling that bones were being reset, resulting in the name "boneset."

Comfrey retained its popularity throughout the nineteenth century, both on the battlefield as well as in domestic settings. Dr. Francis Porcher advocated the herb to the Confederate army for treating malaria due to the scarcity of Peruvian bark, which was used to make quinine. In 1852, Dr. A. Clapp told the American Medical Association that comfrey "deservedly holds a high rank among our indigenous medical plants" (Alfs, 1999, p. 40).

The herb continued to be widely used well into the twentieth century. It enjoyed official pharmaceutical recognition, appearing in the *U.S. Pharmacopoeia* from 1820 to 1916 and the *National Formulary* (1926-1950). Comfrey was particularly popular among the Eclectic physicians during this time. Among the many articles that appeared in *The Eclectic Medical Journal* recommending its use, a 1924 contribution noted: "During the [Spanish influenza] of 1918-19 it was one of the safest and most successful remedies employed and contributed much to the successful management of the disease under Eclectic treatment" (Alfs, 1999, p. 40).

Comfrey experienced a drop in popularity during the mid-twentieth century due to a combination of factors. Its bitter taste did not agree with many users, especially children. The increasing popularity of echinacea and other treatments for the cold also played a role. Of even greater impact was the dramatic rise in the use of aspirin at the turn of the twentieth century. In addition, the declining threat posed by malaria contributed to comfrey's fall from favor. By 1937, the Dispensatory of the United States would note that comfrey was "practically never prescribed by the medical profession and there is no reason for its official recognition" (Alfs, 1999, p. 41).

Since the early 1980s, comfrey has undergone a marked revival in popularity at health food stores in both herbal and homeopathic cold-and-flu preparations. The impetus was provided largely by phytochemists (based mainly in central Europe) who found that a homeopathic preparation of the herb was just as effective as aspirin in treating viral symptoms (e.g., fever, headache, congestion)—this at a time when the negative side effects of aspirin were receiving widespread media coverage. In the mid-1980s, these researchers determined that comfrey's polysaccharides possessed potent immune-stimulating effects which outstripped those found in echinacea. More specifically, a series of immunological tests showed that the plant's water-soluble polysaccharides stimulated human white blood cells to perform more effectively.

Besides health food outlets, comfrey can be obtained in the wild, most notably marshy areas or the perimeters of bodies of water. Picked leaves should be air-dried and then lightly crushed and stored for use as an infusion or liquid extract. While no reports of human toxicity exist in the herbal literature, short-term users are advised to consume it slowly (up to three cups of tea or five droppers of tincture per day). While debate continues as to whether comfrey contains alkaloids (substances present in approximately 15 percent of all vascular plants, including coffee and chocolate), it doesn't appear to contain the toxic pyrrolizidine alkaloids known to have a deleterious effect on the liver in high concentrations. At any rate, the herb's documented effectiveness within a one- to two-day period precludes protracted use.

BIBLIOGRAPHY

Alfs, Matthew. "Boneset." *The Herb Quarterly* (Winter 1999): 40-43.
Gardner, Jo Ann. "Comfrey (Blue Comfrey, Knitbone)." In: *Living with Herbs*. Woodstock, VT: Countryman Press, 1997, pp. 143-145.
Hollis, Sarah. "Comfrey." In: *The Country Diary Herbal*. New York: Henry Holt, 1990, pp. 42-44.
Millspaugh, Charles F. *American Medicinal Plants*. Philadelphia: Lippincott, 1887.

CORIANDER

Sanskrit texts indicate that coriander *(Coriandrum sativum)* was used as early as 5000 B.C. Thebes' Medieval Papyrus, dated around 1550 B.C., refers to the use of coriander seeds at burials. Records also reveal that it was placed in Egyptian tombs during the twenty-first dynasty, between 1091 and 961 B.C. Coriander was ordered to be eaten during Passover in memory of the Hebrews' flight from Egypt. The Chinese, who have cultivated the herb since the fourth century, believed that those eating the seeds during a spiritual trance would achieve immortality. Coriander was also considered to be an aphrodisiac; in some places, it was fed to animals during mating season. In the Middle Ages the seeds were employed to flavor wine, preserves, soups, and meat.

An annual cultivated for at least several thousand years, coriander is commercially harvested today in Europe, the Mediterranean region, China, and the Americas. A thin, spindly-shaped root produces a round, finely grooved stem that matures to about two feet in height. It possesses pinnately decompound leaves and red or white flowers that are arrayed in flat, compound umbels.

The plant's brown, globose seeds release a spicy aroma when opened. The flavor is enhanced further through toasting. Coriander seeds are widely used in breads, stuffing, curried meats, marmalade, and sweets. The herb is also used to flavor bologna and frankfurters.

In Canton, a major urban center located in southeastern China, coriander seeds and leaves are used to neutralize noxious odors emanating from genital areas and halitosis or bad breath.

BIBLIOGRAPHY

Farrell, Kenneth T. "Coriander." In: *Spices, Condiments, and Seasonings.* Westport, CT: AVI, 1985, pp. 92-96.

Heinerman, John. "Coriander *(Coriandrum sativum).*" In: *Heinerman's Encyclopedia of Healing Herbs & Spices.* Paramus, NJ: Prentice-Hall, 1996, pp. 184-185.

Hollis, Sarah. "Coriander." In: *The Country Diary Herbal.* New York: Henry Holt, 1990, pp. 44-45.

Moyler, D.A. "Coriander Seed." In: Charalambous, George (Ed.), *Spices, Herbs and Edible Fungi.* Amsterdam: Elsevier, 1994, pp. 24-27.

Wetherilt, W. and Pala, M. "Herbs and Spices Indigenous to Turkey." In: Charalambous, George (Ed.), *Spices, Herbs and Edible Fungi.* Amsterdam: Elsevier, 1994, pp. 294-296.

CORPORATE ALLIANCE FOR INTEGRATIVE MEDICINE

In October 1998, ten major supplement companies—Natrol, East Earth Herbs, Rexall Sundown, Pure World/Madis Botanicals, Nature's Way, Nature's Herbs, Botanicals International, NuSkin, Nutraceutical Inc., and Weider Nutrition International—united to form the Corporate Alliance for Integrative Medicine. The nonprofit organization possesses a twofold mission:

- To focus on funding research programs at major American universities.
- To develop educational programs regarding supplements directed at medical professionals and consumers.

Given the shared interest of these businesses in creating and marketing new nutritional and medicinal products, herbal advocacy groups had every reason to believe that botanicals would take on an increasingly important role in the twenty-first century.

BIBLIOGRAPHY

"Industry Alliance Formed to Increase Research and Education on Supplements." *Herb World News Online*. Boulder, CO: Herb Research Foundation, 2000. www.herbs.org.

CURRY POWDER

Linguists theorize that "curry" was derived from *kari,* the Tamil (South Indian) word denoting "a sauce." Other sources surmise that it is an English corruption of the shortened Indian word *turri* for the Hindustani *turcurri.* It has evolved into a generic name for virtually any hot spicy dish originating in the East. Although thought to have come from India, the powder mixture is rarely used there. In actuality, premade blends are limited because Indians consider freshly ground spices to be more aromatic and use different proportions of spices for different dishes.

As far back as the third century A.D., the Greeks described curries as Indian cuisine. Kenneth T. Farrell notes, "With the availability of so many native herbs and spices in the various provinces of India for centuries, it is little wonder that the culinary experts of the past experimented with different combinations of these aromatic and flavorful substances to make their limited vegetarian diets more palatable and less monotonous" (p. 228). The higher-quality Indian curries have traditionally blended mixtures of nine to twenty-five herbs and spices, fried together in a hot clarified butter to bring out each of the individual rich bouquets and flavors. (In contrast, curry powders marketed in America are shaken out of a container into a white sauce to accent the main dish.) Although Hindus promoted the use of curries for more than fifteen centuries because the consumption of meat was forbidden, the availability of refrigeration and gradual acceptance of some Western customs in India has resulted in the integration of beef and lamb into present-day offerings.

Curry powder was created for ships and servants of the East India Company returning home. It was intended as a total spice package for a curry; however, recipes vary in keeping with the regional differences within India. Confusion arises from the fact that the powder is now widely incorporated into American and European dishes that have nothing in common with curries. Allowing for these variations, the ingredients comprising curry powder blends include:

- allspice
- anise
- bay leaves
- black pepper
- capsicum
- caraway
- cardamom
- cassia buds
- chili peppers
- cinnamon
- cloves
- coriander
- cumin
- curry leaves
- fennel
- fenugreek
- ginger
- lentils
- mace
- mustard seed
- nutmeg
- paprika
- poppy seed
- saffron
- sesame seed
- turmeric

Ayurvedic healers based in India use curry powder—mixed with water to form a smooth paste—to relieve the itching and inflammation of insect bites and stings. Another alternative healing cure consists of placing a curry blend composed of fresh-ground fenugreek seed and gingerroot, powdered turmeric root, and sesame seed oil.

BIBLIOGRAPHY

Farrell, Kenneth T. "Curry Powder Blend." In: *Spices, Condiments, and Seasonings*. Westport, CT: AVI, 1985, pp. 228-231.

Heinerman, John. "Curry Powder." In: *Heinerman's Encyclopedia of Healing Herbs & Spices*. Paramus, NJ: Prentice-Hall, 1996, pp. 195-197.

D

DIGITALIS

Digitalis *(Digitalis purpurea)* was allegedly discovered by English physician William Withering while searching for an old woman known for her knowledge of medicinal plants as well as the ability to cast spells. Withering was attempting to save a patient on the brink of death from congestive heart failure. According to Heinerman, the physician

> consulted with this old hag, who gave him a curious mixture consisting of a dozen or more herbs and demanded payment. At first he hesitated, believing her price was too high. But when reminded in no uncertain terms that she had the power to put a hex on him if he kept irritating her, the good doctor quickly relented and met her price, leaving the premises with the precious herbs. (p. 203)

Within five days of treatment, the patient was well on his way to recovery. Analyzing the herb mixture in his possession, Withering ascertained that digitalis, also known as foxglove, had been most responsible for the man's recovery. Withering's findings were disclosed in a book titled *An Account of the Foxglove, and Some of Its Medical Uses, with Practical Remarks on Dropsy, and Other Diseases* (Birmingham, England: M. Swinney, 1785). Withering's volume represented the first detailed study of the action of a drug; prior to this, health care givers were limited to drug lists that simply recorded the use of certain drugs for specific conditions without questioning how or why they should work. The publication represented an early step in the development of modern, systematic pharmacology.

Today, digitalis continues to be widely used for cardiac disorders such as atherosclerosis, congestive heart failure, and hypertension. The herb serves to increase the contractibility of heart muscle as well as improving its tone. It also slows the wildly beating ventricles to a normal level by blocking or delaying conduction of the electrical impulse through the atrioventricular node. By improving heart action and blood circulation, digitalis generally improves kidney function, relieves edema, and enables cardiac muscle to compensate for mechanical defects or structural lesions.

Doctors currently prescribe six different glycosides from digitalis: digitalis leaf, digitoxin, digoxin, lanatoside C, acetyldigitoxin, and deslancside. Due to its potency, the herb can only be obtained from a pharmacy with a written prescription.

BIBLIOGRAPHY

Garland, Sarah. *The Complete Book of Herbs & Spices*. Pleasantville, NY: Reader's Digest Association, 1993.

Heinerman, John. "Digitalis *(Digitalis purpurea)*." In: *Heinerman's Encyclopedia of Healing Herbs & Spices*. Paramus, NJ: Prentice-Hall, 1996, pp. 203-204.

DILL

Dill is a derivation of the Norwegian word "dilla," meaning to soothe. The Babylonians and Assyrians cultivated it during antiquity for its magical and medicinal powers. Its seeds have been referred to as Meeting House Seeds because when brought to church many generations past, the congregation would nibble on them during prolonged sermons.

An annual typically reaching as much as two-and-one-half feet in height, dill *(Anethum graveolens)* is distinctively aromatic in character, a milder and sweeter variant of caraway. Its taste is reminiscent of fennel, though somewhat more pungent and aggressive in flavor. Indigenous to France, Spain, and Russia, it is presently cultivated in North America, much of Europe, India, and Pakistan.

Dill's long, spindle-shaped root supports a smooth, shiny, and hollow stem. It has feathery leaves, accented by linear and pointed leaflets. In midsummer, the plant bears flat terminal umbels with an abundance of yellow flowers. Its flat fruit is produced in prodigious quantities; one ounce contains more than 25,000 seeds.

The dill seed is employed worldwide as a flavoring agent. It is especially popular with pickled cucumbers and beets, beans, cabbage, cauliflower, cottage cheese, fish, hors d'oeuvres, green apple pies, peas, and soups.

The seeds are also valued in a variety of medicinal applications. Some cultures chew them as a means of combating halitosis. They are combined with warm white wine as a drug-free remedy for sleeplessness. A tea made of white wine and a blend of anise, coriander, caraway, and dill seeds is reputed to stimulate the flow of breast milk in nursing mothers. In recognition of its widespread popularity, McCormick's *Flavor Trends 2000* designated it a "Top 10 Flavor for 2000."

Rodent research findings published in 1984 found that limonene, an essential oil found in dill, caraway, and celery seed oils as well as in citrus fruit peel oils, could be responsible for reactivating a natural anticarcinogenic process. While findings remain inconclusive thus far, the latter half of the 1990s saw clinical research concerning the role of limonene in relation to human pancreatic and colorectal cancers.

BIBLIOGRAPHY

Farrell, Kenneth T. "Dillseed—Dillweed." In: *Spices, Condiments, and Seasonings*. Westport, CT: AVI, 1985, pp. 101-105.

Gardner, Jo Ann. "Dill (Dill Weed)." In: *Living With Herbs*. Woodstock, VT: Countryman Press, 1997, pp. 156-168.

Heinerman, John. "Dill *(Anethum graveolens)*." In: *Heinerman's Encyclopedia of Healing Herbs & Spices*. Paramus, NJ: Prentice Hall, 1996, pp. 204-206.

E

ECHINACEA

Commonly known as purple coneflower, echinacea encompasses nine different species, three of which are utilized commercially: *Echinacea angustifolia, Echinacea purpurea,* and *Echinacea pallida*. *E. purpurea* can be grown with a minimum of expertise and is, therefore, found in many herb gardens. *E. angustifolia,* on the other hand, is not easy to cultivate, and the widespread demand for the plant has severely diminished its availability in the wild. Although useful, *E. pallida* is believed to have less medicinal value than the aforementioned species.

Echinacea possesses distinctive daisy-like purple flowers with a pronounced cone-shaped center. All parts of the plant have medicinal value, but the root is its most potent portion. The root is most effective when three to five years in age; treatment is generally short-term (less than a month in duration). Echinacea's dried, aerial portions are best employed as a tonic, and can be taken for longer periods. Its active constituents include volatile oil, glycosides, amides, antibiotic polyacetylenes, and insulin.

Enchinacea is native to the U.S. plains; according to evidence culled from archeological sites, Native Americans used it as at least as far back as the 1600s for treating a wide range of conditions, including colds, flu, skin infections, gangrenous wounds, toothaches, gonorrhea, cramps, hydrophobia, scarlet fever, and snakebites. In fact, in his book, *Echinacea: The Immune Herb!,* Christopher Hobbs states that "the Plains Indians revered echinacea above all other herbs" (Langer, 2000). The European settlers learned about the herb from various tribes and utilized it in much the same manner.

Around 1870, German physician H. C. F. Meyer developed a patent medicine containing echinacea. He actively promoted "Meyer's Blood Purifier" with both Eclectic and mainstream doctors. As a result, echinacea was introduced into the materia medica by 1887 and had become the most popular herb in medical practice by 1907. Despite controversy surrounding its use (the American Medical Association has never officially accepted it), echinacea's popularity—aided by the support of homeopathic physicians—spread to Europe early in the twentieth century. More than 400 scientific journal articles appeared between the 1930s and 1980s exploring its medicinal properties. By 1986, more than 240 medicinal products with echinacea as an ingredient were available in Germany alone, providing the impetus for U.S. herbalists to "rediscover" the plant.

During the 1990s, echinacea's popularity grew to even greater proportions; it is now found in tinctures, glycerites, creams, throat sprays, lozenges, mouthwash, and capsules. According to Hobbs, it is used for general infections and wound healing, colds, flu, candidiasis, strep throat, staph infections, urinary tract infections, pelvic inflammatory disease, tonsil and non-strep throat infections, upper respiratory tract infections, burns, herpes, skin ulcers, psoriasis, whooping cough, bronchitis, leukopenia, rheumatoid arthritis, allergies, toothaches, mouth and gum infections, bites and stings, blood and food poisoning, boils and carbuncles, and eczema. And that's not all! Although not yet conclusive, studies indicate that it may be effective in controlling AIDS. Furthermore, echinacea use has revealed few side effects; high doses may cause nausea or dizziness and patients with autoimmune conditions such as lupus or AIDS are potentially at risk.

BIBLIOGRAPHY

Griggs, Barbara. *Green Pharmacy: The History and Evolution of Western Herbal Medicine.* Rochester, VT: Healing Arts, 1997.

Hobbs, Christopher. *Echinacea: The Immune Herb! Second Edition.* Santa Cruz, CA: Botanica Press, 1995.

Langer, Robin E. "Herb-of-the-Month: Echinacea sp.—November 2000." *Snowbound Herbals* (November 4, 2000): 2p. www.sbherbals.com.

EUCALYPTUS

One of the fastest growing and largest trees in existence, Eucalyptus *(Eucalyptus globulus)* can reach over 250 feet in height. Its extensive root network has rendered it useful in draining marshy areas in hot, humid climates. Guatemala adopted a program of planting eucalyptus in order to control the spread of malaria. Native to Australia, the tree is now cultivated worldwide.

The tree possesses bluish-white, peeling bark and green branchlets. Leaves are opposite, heart-shaped, bluish-green, and sticky when young; however, mature leaves are alternative, lance-shaped, green, and smooth in appearance. They emit a distinctively pungent odor when crushed.

The eucalyptus figures prominently in the most famous Australian song of all time, "Waltzing Matilda." Composed in 1895 by a bush poet, Andrew Barton Paterson, to memorialize the efforts of workers who stood up for their rights against powerful wool industry landlords, the last verse depicts a swagman "drowning himself by the Coolibah [eucalyptus] tree" in order to evade capture by police.

Various biographies of Paterson indicate that he was well aware of the medicinal value of the Coolibah tree. In one instance, the poet rubbed eucalyptus oil on his gum lining to combat an infection. Bothered by a loose molar on another occasion, he placed a piece of cloth soaked in the oil between his cheek pouch and the tooth; after several days, the gum lining around the tooth tightened up and the problem disappeared for good.

Paterson described rubbing eucalyptus oil over the exposed parts of his body to repel annoying "biting back flies." One historian tried the remedy and was "very pleased to discover just how well it worked to keep the little bastards away from me." Paterson also added a few drops of the oil to whiskey, brandy, or hot water to remove mucus accumulation from his lungs, control coughing, clear up fever, and relieve the laryngitis that sometimes resulted from singing too long with his banjo to audiences.

Other aboriginal applications for eucalyptus oil have included treating stomach disorders, liver and bladder infections, wounds, and sores. Added to tepid bath water, it relieves arthritis, lower backache, lumbago, and rheumatism.

Modern medicinal manufacturers have also found eucalyptus to be a useful ingredient. A popular over-the-counter skin ointment, Vicks VapoRub, has been popular for decades in the treatment of respiratory ailments, most notably asthma and bronchitis. The Halls Mentho-Lyptus brand of cough drops contains 12 percent eucalyptus oil, peppermint oil, and other herbs.

BIBLIOGRAPHY

Heinerman, John. "Eucalyptus *(Eucalyptus globulus)*." In: *Heinerman's Encyclopedia of Healing Herbs & Spices*. Paramus, NJ: Prentice-Hall, 1996, pp. 222-225.

McHoy, Peter, and Westland, Pamela. *The Herb Bible*. New York: Barnes & Noble Books, 1994.

EVENING PRIMROSE

Evening primrose *(Oenothera biennis)* is known by a variety of common names (which allude either to its resemblance to other plants or alleged medicinal properties): German rampion, king's cure-all, night willow herb, and scabish. A biennial that matures to eight feet in height and flowers in its second season, the herb's family members include fireweed and fuchsia. Considered a weed in Europe, evening primrose is native to North America; it is highly visible in backyards, fields, roadsides, and waste areas.

Because all parts of the plant are edible, evening primrose has had a long history as a food staple. Native Americans based in Utah and Nevada are known to have eaten the seeds. They are sometimes used as a substitute for poppy seeds, to which they bear a physical resemblance. Young leaves have been served raw in salads or as a potherb; however, they are generally cooked in several changes of water to expunge any lingering bitterness. English colonists brought the seeds from North America back to the British Isles as early as 1614; from that time onward, evening primrose was cultivated in the gardens of England, Germany, and other European countries for its nut-flavored roots (typically boiled like parsnips). In the mid-nineteenth century, it was reintroduced to the New World as a vegetable called German rampion due to its similarity to rampion, a bellflower possessing edible roots and basal leaves.

Native Americans also pioneered medicinal uses of evening primrose. The Ojibwa made a poultice from the entire plant for treating bruises, and the Cherokee brewed a tea from the root to stimulate weight loss. By the eighteenth century, European colonists were using evening primrose as medicine. The Shakers used leaves or roots externally to facilitate the healing of wounds as well as for making a tea to settle an upset stomach.

The plant is especially popular today for controlling discomfort associated with premenstrual syndrome (PMS). A number of studies have found that symptoms such as irritability, breast pain and tenderness, and mood changes are significantly reduced by the use of evening primrose oil.

During the 1990s, evening primrose received increasing attention as a dietary supplement and a valuable source of essential fatty acids. The latter—including cis-linoleic acid, alpha-linolenic acid, and gamma-linolenic acid (GLA)—are substances the body cannot manufacture but requires for maintenance of health.

Many countries have formally approved various dietary and medicinal uses of the plant. In the United States, the Dietary Supplement Health and Education Act of 1994 has classified evening primrose oil as a dietary supplement. In 1995, Steven Foster noted that the evidence gathered from

> more than 120 studies in university hospitals in fifteen countries . . . suggests that evening primrose oil may be useful for treating a wide range of conditions associated with imbalances and abnormalities of essential fatty acids, including atopic eczema, asthma, migraine, inflammations, PMS, metabolic disorders, diabetes, arthritis, and alcoholism. (p. 66)

BIBLIOGRAPHY

Briggs, C.J. "Evening Primrose." *Canadian Pharmaceutical Journal 250* (1986): 248-254.

Foster, Steven. "Evening Primrose: An Important Treatment for Essential-Fatty-Acid Deficiency." *The Herb Companion* (October/November 1995): 65-66.

Garland, Sarah. *The Complete Book of Herbs & Spices*. Pleasantville, NY: Reader's Digest Association, 1993.

Olin, B.R. (Ed.). "Evening Primrose." *Lawrence Review of Natural Products* (November 1995).

F

FEDERAL REGULATION OF HERBS AND SPICES

Laws Regulating Herbs and Dietary Supplements

Food, Drug, and Cosmetic Act

In 1938, Congress passed the federal Food, Drug, and Cosmetic Act, which served to:

- extend control to cosmetics and therapeutic devices;
- require new drugs to be shown safe before marketing;
- institute a new system of drug regulation;
- eliminate the Sherley Amendment requirement to prove intent to defraud in drug misbranding cases;
- provide that safe tolerances be set for unavoidable poisonous substances;
- authorize standards of identity, quality, and fill-of-container for foods;
- authorize factory inspections; and
- add the remedy of court injunctions to the previous penalties of seizures and prosecutions.

The Food Additives Amendment was added in 1958, requiring manufacturers of new food additives to establish safety. The Delaney proviso prohibited the approval of any food additive shown to induce cancer in humans or animals, and the FDA published the first list of substances, including herbs and spices, as Generally Recognized As Safe (GRAS) in the *Federal Register.* The list contained nearly 200 substances. This list is now compiled in Title 21 of the *Code of Federal Regulations.*

Nutrition Labeling and Education Act of 1990 (NLEA)

The Nutrition Labeling and Education Act of 1990 requires packaged foods to bear nutrition labeling and health claims that are consistent with terms defined by the Secretary of Health and Human Services (i.e., approved by the FDA). The law preempts state requirements regarding food standards, nutrition labeling, and health claims, and, for the first time, authorizes some health claims for foods. The food ingredient information panel, serving sizes, and terms such as "low fat" and "light" are standardized. The phrase "herbs, or similar nutritional substances" was added to the term "dietary supplement."

The NLEA has been successful in raising consumer awareness and confidence about the role of nutrition in promoting health and reducing the risk of disease, and has brought structure to a field that in the 1980s had no regulatory framework. With this success and the growing interest in health maintenance promotion and disease risk reduction, NLEA served as a springboard to the Dietary Supplement Health and Education Act (DSHEA) of 1994, which established the legal basis for dietary supplements to become a core element of self care in the United States.

Dietary Supplement Health and Education Act of 1994 (DSHEA)

In 1994, DSHEA defined dietary supplements and dietary ingredients, established a new framework for assuring safety, outlined guidelines for literature displayed where supplements are sold, provided for the use of claims and nutritional support statements, required ingredient and nutrition labeling, and granted the FDA the authority to establish good manufacturing practice regulations. The law also required formation of an executive level Commission on Dietary Supplement Labels and an Office of Dietary Supplements within the National Institutes of Health. Enforcement authority for the FDA under the Food, Drug, and Cosmetic Act and of the Federal Trade Commission (FTC) under the Federal Trade Commission Act remained unaltered and substantial.

Food and Drug Administration Modernization Act (FDAMA)

Before the Food and Drug Administration Modernization Act of 1997, companies could not use a health claim or nutrient content claim in food labeling unless the FDA published a regulation authorizing such a claim. Two new provisions of FDAMA now permit distributors and manufacturers to use such claims if they are based on current published authoritative statements from certain federal scientific bodies, or from the National Academy of Sciences. These provisions are intended to expedite the process by which the scientific basis for such claims is established, and they place dietary supplements on an equal footing with conventional foods.

Regulation Myths

Despite popular beliefs from consumer advocates and from the media about an "unregulated industry," the herb and dietary supplement industry, which includes spices, is one of the best regulated with a substantial number of laws, regulations, and proposed regulations at the federal level. According to remarks by FDA Commissioner Jane E. Henney before the House Committee on Government Reform, March 25, 1999, "FDA has tools at its disposal to take enforcement actions against dietary supplements found to have safety, labeling, or other violations of the FD+C Act, as amended by the Dietary Supplement Health and Education Act (DSHEA) of 1994" (Soller, 2000, p. 67).

Both the Food and Drug Administration (FDA) and the Federal Trade Commission (FTC) have displayed considerable enforcement authority, since DSHEA, to ensure that safe, beneficial, and quality dietary supplements are available for health promotion and maintenance and disease risk reduction, as intended by Congress in the passage of DSHEA. Despite assertions to the contrary usually made by persons unfamiliar with regulatory issues, the herb and dietary supplement industries are substantially regulated; however, these regulations are significantly different from those in the pharmaceutical industry. The FDA has to prove that the herbs and spices are unsafe before they can regulate. Spices and flavorings are used in foods and alcoholic beverages, both of which are regulated for filth, identification, decomposition, and other components; the list is then published as part of GRAS. Herbs used in pesticides are regulated, but until it can be established that herbs are foods, it will be difficult to ascertain that they are capable of killing people.

Much of the confusion surrounding the regulation of herbal products has abated since the passage of DSHEA. No other law has ever received as much direct grassroots advocacy, with the number of supportive letters to Congress topping two million. Congressional intent in passing the law was to give consumers the right to make their own choices about "preventive health care programs based on data from scientific studies of health benefits related to particular dietary supplements" (Cichowicz interview, 2000).

FDA and FTC

The FDA and the FTC have a long-standing agreement governing the division of responsibilities between the two agencies. With regard to dietary supplements, the FDA has the power to:

- stop any company from selling a dietary supplement that is toxic or unsanitary;
- stop the sale of a dietary supplement that has false or unsubstantiated claims;
- take action against dietary supplements that pose "a significant unreasonable risk of illness or injury";

- stop any company from making a claim that a product cures or treats a disease;
- stop a new dietary ingredient from being marketed if the FDA does not receive enough safety data in advance; and
- require dietary supplements to meet strict manufacturing requirements (Good Manufacturing Practices), including potency, cleanliness, and stability.

The FTC has the power to:

- enforce laws outlawing "unfair, or deceptive acts or practices" to ensure consumers get accurate information about dietary supplements, so they can make informed decisions about these products;
- challenge and stop advertising that is not adequately substantiated;
- investigate complaints or questionable trade practices either informally or formally, where it has strong compulsory investigative authority, including the power to require a respondent to produce documents, give testimony, or answer written questions;
- negotiate a consent order or proceed through an FTC adjudication resulting in a cease and desist order, which can be quite broad in its scope;
- seek preliminary and permanent injunctions to stop false advertisements or other violations of the FTC Act; and
- seek civil penalties for violations of trade regulation rules.

Since the passage of DSHEA, both the FDA and the FTC have been active in building the regulatory framework for dietary supplements. For example, in the last four decades, the FDA has published 143 proposed and final rules in the *Federal Register* relating to dietary supplements. Between 1962 and 1981, the FDA issued fifty-two such publications, most related to its attempt to regulate vitamins and minerals as drug products. From December 1981 to May 1993, the FDA issued no documents pertaining to regulation of dietary supplements, but since the passage of DSHEA in 1994, the agency has issued ninety-one *Federal Register* publications, a 66 percent increase since 1962. These proposed and final rules cover such important enforcement topics as labeling, good manufacturing practices, development of a comprehensive regulatory strategy for dietary supplements, and issues relating to ingredient safety.

BIBLIOGRAPHY

Cichowicz, Stanley, FDA Botanist. Interview with Martin Manning, December 12, 2000.

Slevin, Peter. "Another Pardon Stirs Controversy." *The Washington Post,* February 6, 2001, pp. A3, A10.

Soller, R. William. "Regulation in the Herb Market: The Myth of the Unregulated Industry." *HerbalGram 49* (2000): 64-67.

FENNEL

Fennel *(Foeniculum vulgare),* a perennial frequently growing to more than five feet in height, possesses long, feathery leaves, smooth erect stems, expansive umbels of yellow flowers (which bloom in midsummer), and broad, aromatic seeds. Native to the Mediterranean region, it was introduced to northern Europe by the Romans, and to North America by Old World colonists. It thrives in the wild, particularly on dry roadsides and near the sea.

The leaves and stems are often picked fresh during the summer; the latter can also be dried prior to use as a flavoring agent. The seed heads are collected when the seeds mature and hung in an airy place to dry. The seeds possess a slightly bitter, aniseed flavor, whereas the leaves are endowed with a sweeter, richer taste.

Fennel has a long history of decorative, culinary, medicinal, and magical uses. The Greeks named it *Marathron,* from *maraino,* to grow thin. William Coles, the seventeenth-century English botanist, noted that the herb was widely used "for those that are grown fat, to abate their unwieldiness and cause them to grow more gaunt and lank."

The Romans, Chinese, and Hindus regarded fennel as an antidote for snakebite and other poisons. The Romans, in particular, valued the herb as a food and medicine. The natural historian, Pliny, listed it as a treatment for twenty-two different ailments; furthermore, it was purported to increase the flow of mothers' milk, to improve one's strength and courage, and to prolong youthful vigor.

Following the Roman occupation, the English considered fennel to be of value not only as a medicine, but as protection against all evil. The ancient *Nine Herbs Charm* noted the herb's power: "It stands against pain, resists the venom, It has the power against three and against thirty, Against a fiend's hand and against trick, Against witchcraft of vile creatures."

During the Middle Ages, fennel was studded in keyholes and hung over doors to protect against witches, particularly on Midsummer Eve. The seeds were also consumed in large quantities with both fish and hard fruit as an aid to digestion.

The herb is best known today as a seasoning for fatty meats (e.g., pork) and oily fish as well as chicken, ham, lamb, fresh vegetables, and creamy sauces. The seeds are used to flavor bread and cakes, while the fresh leaves function as a mint alternative in yogurt or hummus. Salads are enhanced by the leaves, peeled stalks, and bulbs; the latter can also be braised as a vegetable. Fennel's popularity is such that McCormick's *Flavor Trends 2000* included the herb in its "Top 10 Flavors for 2000."

The oil derived from fennel seeds is recognized today for its stimulant and digestive properties. The leaves are used as a cleanser and astringent for the skin. Tired or inflamed eyes are often treated by a strained decoction consisting of water boiled with fennel leaves. This practice also goes back to antiquity. Pliny tells of snakes eating and rubbing against the herb in order to restore and sharpen their sight after getting rid of old skins. The herbalist Culpeper states that the fresh juice "dropped into the eyes cleanses them from mists and films that hinder the sight."

BIBLIOGRAPHY

Farrell, Kenneth T. "Fennel Seed." In: *Spices, Condiments, and Seasonings*. Westport, CT: AVI, 1985, pp. 106-109.

Garland, Sarah. *The Complete Book of Herbs & Spices*. Pleasantville, NY: Reader's Digest, 1993.

Hollis, Sarah. "Fennel." In: *The Country Diary Herbal*. New York: Henry Holt, 1990, pp. 54-55.

FENUGREEK

An erect annual reaching approximately two feet in height, fenugreek *(Trigonella foenumgraecum)* is similar in habit to lucerne hay. The genus name is Greek in derivation, denoting "three-angled," in reference to the plant's corolla. The species name means "Greek hay"; in the past, the plant was used to scent inferior hay.

Fenugreek has been grown since antiquity in India, western Asia, and the Nile Valley for its medicinal value and as spring forage. It is presently cultivated in Algeria, Argentina, Cyprus, Egypt (going back as far as 1000 B.C.), France, Germany, Greece, India, Italy, Lebanon, Morocco, Portugal, Spain, Yugoslavia, and the United States.

The brownish seeds are one-half inch in length, oblong, possessing a deep furrow dividing them into unequal lobes. They are grouped (ten to twenty together) in long, narrow, sickle-like pods.

Fenugreek is extremely popular as a flavor enhancement, including in salads and various curry blends. The powdered seeds are added to a wide range of Eastern foods, most notably in India and Pakistan, to increase tartness.

Based on research in Villemoisson-sur-Orge, France, in which mongrel dogs were given a standard diet supplemented with fenugreek seed meal for eight weeks, the plant appears to play a role in reducing cholesterol. It is now available in capsule form for general use (e.g., Cholester-Low, manufactured by the St. Petersburg-based company, Old Amish).

Fenugreek can also be brewed as a tea to treat hay fever attacks and ringing in the ear. Because boiling the seeds can result in an extremely bitter tea, John Heinerman recommends soaking them in cold water overnight and drinking the strained liquid (unheated) the next day. The process can be repeated for several days before the seeds need to be replaced.

BIBLIOGRAPHY

Farrell, Kenneth T. "Fenugreek Seed." In: *Spices, Condiments, and Seasonings.* Westport, CT: AVI, 1985, pp. 110-113.

"Fenugreek (Cemen Otu)." In: Charalambous, George (Ed.), *Spices, Herbs and Edible Fungi.* Amsterdam: Elsevier, 1994, pp. 301-303.

Heinerman, John. "Fenugreek *(Trigonella foenumgraecum).*" In: *Heinerman's Encyclopedia of Healing Herbs & Spices.* Paramus, NJ: Prentice-Hall, 1996, pp. 232-234.

FEVERFEW

The medicinal benefits of feverfew *(Tanacetum parthenium)* have been widely known as far back as antiquity. Dioscorides, a Greek physician serving in the Roman army in the first century, stated in his materia medica that the herb was effective in dealing with "melancholicall" (or "headaches," in modern terms).

The plant remained popular during the early modern era. Gerard's *Herball,* originally produced in 1597, stated:

> Feverfew dried and made into pouder, and two drams of it taken with honie or sweet wine, purgeth by siege melancholy and flegme, whereforre it is very good for them that are giddie in the head, or which have the turning called *Vertigo,* this is a swimming and turning in the head. Also it is good for such as be melancholike, sad, pensive, and without speech.

English Physician Enlarged, written by Nicholas Culpeper in the seventeenth century and reputed to be the most widely published English-language herbal ever, noted, "It is very effectual for all pains in the head coming of a cold cause, the herb being bruised and applied to the crown of the head; as also for the Vertigo, that is a sunning or swimming of the head."

In recent years, a series of experiments have provided added insights regarding the efficacy of feverfew. A 1985 study conducted by E. S. Johnson at the City of London Migraine Clinic in collaboration with the Chelsea College of the University of London, tested human volunteers already using the herb. The group that was administered a placebo reported a significant increase in the frequency and severity of migraines (and associated symptoms such as nausea and vomiting), whereas the group continuing to ingest feverfew claimed no change in their migraines. A randomized, double-blind, placebo-controlled trial in 1988 reinforced the Johnson study.

These experiments provided evidence that migraine attacks cause blood platelets to release the hormone serotonin, which constricts blood vessels. The parthenolide content in the dried feverfew leaf was found to inhibit the release of serotonin, thereby suppressing migraine symptoms. The publication of these findings in the scientific and popular literature led to a proliferation of feverfew-based products in North America by the early 1990s. Early reviews were mixed, primarily because these medications all contained less than the minimum 0.2 percent parthenolide per daily dosage recommended by D. V. C. Awang and his colleagues in a 1991 report published in the *Journal of Natural Products,* a figure later adopted by Canadian regulatory authorities (cited in Foster, 1995, p. 65). The latest wave of products, however, have included a higher parthenolide content.

For those seeking an alternative to commercial products, feverfew is easy to cultivate and prepare fresh. Modern herbals typically recommend eating two to four fresh leaves daily to treat migraines. Only one drawback to this procedure has come to light; approximately 11 percent of those using fresh leaves have reported mouth ulcers. Feverfew's popularity is reflected by its inclusion in Discount Natural Herbs' "Top Ten Selling Herbs" through mid-2001.

BIBLIOGRAPHY

Foster, Steven. "Migraine Headaches and Feverfew." *The Herb Companion I*(4) (October/November 1995): 63-65.

Gardner, Jo Ann. "Feverfew." In: *Living with Herbs*. Woodstock, VT: Countryman Press, 1997, pp. 163-164.

Garland, Sarah. *The Complete Book of Herbs & Spices*. Pleasantville, NY: Reader's Digest Association, 1993.

FLAX

Common flax *(Linum usitatissimum)* has been cultivated by humankind as long as any member of the plant kingdom. It is closely related to approximately 100 other flax species, most notably mountain flax, perennial flax, and prairie flax. It has had an extensive array of uses, including food, clothing (linen), medication, and industrial products.

Extensive evidence is available regarding the importance of flax to prehistoric societies. The presence of linen wrapped around a tool handle at an archeological excavation of a small village near the headwaters of the Tigris River in what is now Turkey is believed to be the oldest piece of cloth (dated 7000 B.C.) ever discovered. Similar remnants have also been located in Stone Age Swiss villages.

Since the advent of recorded history, flax has remained a vital component of civilization. Archeological digs and tablet inscriptions at Pylos indicate flax was cultivated in early Greek societies (1900-1700 B.C.). Homer notes that flax was instrumental in making clothes, fishing nets, and ship sails. The Old Testament contains many references to flax and the fine linens produced from its fibers; the New Testament states that Christ was wrapped in a linen shroud during interment.

The plant also enjoys a rich mythological heritage. Bohemian folklore relates that seven-year-old children were encouraged to dance in flax fields in order to attain physical beauty. Citing Teutonic myths, Dr. William Haggerty relates that

> flax flowers were believed to serve as protection from sorcery under the watch of the goddess Queen Hulda, who lived in a secret cave in the mountains. It was she who taught mortals to grow, spin, and weave flax fibers. Twice each year she visited her domain, once in the summer when the blue flax flowers brightened the fields, and again during the 12 nights before the feast of the Epiphany, when all the gods and goddesses were believed to visit the earth. (p. 51)

J. W. Krutch has argued that flaxseed was employed as a food even before early cultures learned to convert its fibers to clothing. It possesses a high nutritional value and has long been favored in cereal and bakery products. Linseed cake—a by-product after the expressed oil has been utilized as an intermediate for alkyd resins, automobile brake linings, fabric treatments, fine chemicals, linoleum, nonvolatile organic coatings, paint formulations, and printing inks—serves as animal feed.

Likewise, flaxseed has been used medicinally for a variety of ailments. According to Krutch, first-century Greek physician Dioscorides recommended it for "mollifying all inflammation." External doses consisted of flaxseed boiled with honey, oil, and a little water; internal doses were limited to boiled honey and flaxseed. Dioscorides also touted a poultice prepared from flaxseed, honey, and figs to treat sunburn as well as one mixing the seed with honey and pepper for enhancing sexual pleasure. In the seventeenth century, English herbalist Nicholas Culpeper related that "the seed being boiled in water and applied as a poultice or plaster assuageth all pains, softeneth cold, tumors or swellings, the impostumes of the neck and other parts of the body . . . pounded with figs, it is good to ripen and bring to a head boils and other swellings."

K. Kindscher notes that blue flax was used for eyewashes by the Paiutes and Shoshones of the Great Plain Basin. Preparations from common flax cultivated in the region were also employed by settlers as folk remedies for coughs, gallstones, and constipation in addition to digestive, urinary, and lung disorders.

Preparations based on the herb remain very popular today. D. Bown, author of *The Herb Society of America Encyclopedia of Herbs and Their Uses,* recommends flaxseed for constipation, coughs, gastritis, hardening of the arteries, pharyngitis, rheumatoid arthritis, and sore throats. She also notes its use externally for abscesses, boils, burns, scalds, sores, and ulcers, and as a poultice (mixed with white mustard seeds) to treat chest complaints. Linseed oil was included in the *United States Pharmacopeia* and *National Formulary* from 1820 to 1947. It has been covered in many European pharmacopeias, including those of France, Germany, and the United Kingdom.

Haggerty cites the following factors as causing flaxseed's beneficial nutraceutical effects:

- Its oil contains high levels (53 percent) of a-linolenic acid,
- Flaxseed fiber contains high levels of a lignan called secoisolariciresinol diglucoside, and
- Flaxseed contains certain beneficial gums and mucilage.

A significant number of tests have been conducted in recent years regarding the relationship of these characteristics to the prevention and/or treatment of cancer, systemic lupus erythematosus, hyperlipidemia (high cholesterol), malaria, and rheumatoid arthritis. While many studies remain inconclusive, findings have often been promising.

BIBLIOGRAPHY

Bown, Deni. *The Herb Society of America Encyclopedia of Herbs and Their Uses.* London: Dorling Kindersley, 1995.

Cubbabe, S.C., and Thompson, L.U. (Eds.). *Flaxseed in Human Nutrition.* Champaign, IL: AOCS Press, 1995.

Garland, Sarah. *The Complete Book of Herbs & Spices.* Pleasantville, NY: Reader's Digest Association, 1993.

Haggerty, William. "Flax: Ancient Herb and Modern Medicine." *HerbalGram 45* (1999): 51, 57. Also available online at www.AllHerb.com.

Hall, A.V., Parbtani, A., Clark, W.F., Spanner, E., Keeney, M., Chin Yee, I., Philbrick, D.J., Holob, B.J. "Abrogation of MRL/LIPR Lupus Nephritis by Dietary Flaxseed." *American Journal of Kidney Disease 22* (1990): 326ff.

Hu, F. "Flaxseed in Human Nutrition." American Dietetic Association's 81st Annual Meeting, Kansas City, MO, October 20, 1998.

Kindscher, K. *Medicinal Wild Plants of the Prairie: An Ethnobotanical Guide.* Lawrence, KS: University of Kansas Press, 1992.

Krutch, J.W. *Herbal.* New York: G.P. Putnam's Sons, 1965.

Thompson, L.U., Seidt, M.M., Rickard, S.F., Orcheson, L.J., Fong, H.H. "Antitumorigenic Effect of a Mammalian Lignan Precursor from Flaxseed." *Nutrition and Cancer 26* (1996): 159ff.

FOLEXCO

The Montgomeryville, Pennsylvania-based Folexco initially made its name as a producer of extracts and flavors. It is affiliated with the A. M. Todd Company, based in Kalamazoo, Michigan, which has annual sales of approximately $300 million.

On April 23, 1999, the company acquired a controlling percentage of East Earth Herb stock, making it the leading U.S. botanical extract manufacturer. Headquartered in Eugene, Oregon, East Earth Herb began supplying botanical extracts to the dietary supplement industry in the early 1970s. At the time of the takeover, it was reporting annual sales of fifteen million dollars.

BIBLIOGRAPHY

"Folexco and East Earth Herb Form Largest Botanical Extract Manufacturer in US." *Herb World News Online.* Boulder, CO: Herb Research Foundation, 2000. www.herbs.org.

FOOD PREPARATION AND NUTRITION: HERBS, SPICES, AND BOTANICALS

The use of herbs, spices, and related botanicals has been vital to the preparation of meals throughout history. This importance is reflected in the diversity of traditions and customs surrounding their use with certain foods. Certain recipes are closely identified with one nation or culture; for example, spicy curries with India. On the other hand, individual preferences are much harder to gauge. Sarah Garland notes that "garlic is an obvious example of a herb that some people find offensive in the smallest amounts, while others will greedily eat whole bulbs simmered in an eel sauce or pounded in aioli" (Garland, 1993, p. 158).

Herbs and spices cut across all major food groupings, including

- Seasonings: salts, peppers, spice mixtures (e.g., French spice mixture, garam masala, five-spice powder, chili powder), herb mixtures (e.g., fines herbes, bouquet garni), mustards
- Marinades
- Oils, vinegars, and dressings
- Sauces
- Hors d'oeuvres
- Drinks
- Soups
- Salads
- Relishes
- Dairy products
- Main dishes
- Desserts and other sweets

Many herbs and spices are not only rich in vitamins and minerals, but benefit health in additional ways. For instance, cinnamon is a stimulant, a digestive aid, and an antiseptic.

BIBLIOGRAPHY

Farrell, Kenneth T. *Spices, Condiments, and Seasonings*. Westport, CT: AVI, 1985.
Flavor Trends 2000. Baltimore, MD: McCormick & Company, 2000.
Garland, Sarah. *The Complete Book of Herbs & Spices*. Pleasantville, NY: Reader's Digest Association, 1993, pp. 158-219.

FRANKINCENSE

As with myrrh, Christian scriptures (i.e., gift of the Magi to the baby Jesus) have conferred legendary status upon frankincense *(Boswellia thurifera)*. Both spices also have similar origins; they are prepared from lumps of light reddish to yellow-brown gum secreted by several tree species.

Sometimes called olibanum, frankincense was extremely popular with ancient Egyptians as an embalming and fumigating agent. It is still employed in the Near East as a masticatory, or mouth cleanser. Its oil extract remains in demand for high-quality perfumes, most notably of the oriental and floral variety.

The spice is also used to treat a number of mental and physical conditions. Physicians at one infirmary have claimed that a steamy concoction of equal parts frankincense resin, whole cloves, and fresh cardamom played a role in significantly improving the memory of 63 percent of their patients. In *Heinerman's Encyclopedia of Healing Herbs & Spices,* Amra Kevic tells of the practices of Yugoslavian smugglers who were bringing back contraband purchased at Turkish bazaars—including frankincense—aboard the Istanbul-Belgrade train line. After a series of harrowing adventures with a friendly Yugoslavian peasant woman, which included being thrown off the train for refusing to pay a black market assessment, Kevic was invited on a tour of Belgrade's hospitals specializing in respiratory disorders in order to observe the benefits of such clandestine activities:

> There she was shown by members of the medical staff how pieces of the frankincense were dissolved in hot water and pa-

tients suffering from asthma, bronchitis and emphysema were told to inhale its warm vapors. (p. 241)

BIBLIOGRAPHY

Heinerman, John. "Frankincense *(Boswellia thurifera)*." In: *Heinerman's Encyclopedia of Healing Herbs & Spices*. Paramus, NJ: Prentice-Hall, 1996, pp. 240-241.

McHoy, Peter, and Westland, Pamela. *The Herb Bible*. New York: Barnes & Noble Books, 1994.

G

GARLIC

Garlic *(Allium sativum)* is one of the most versatile herbs, recognized worldwide for its culinary, medicinal, and cosmetic uses. A hardy perennial that responds best to full sun and fertile loam, garlic is propagated by early spring or late fall planting of segments of the bulb (known as "cloves," each of which is sheathed in a papery skin, and held together by a tissue-like wrapping). The mature plant has flat gray-green leaves about one inch wide and twelve inches long. Willowy flower stalks produce clusters of mauvish-white flowers that can reach a length of two or more feet.

Garlic has been popular for centuries as a seasoning for foods, most notably potatoes, pasta dishes, soups, salads, meats, and poultry. With its pronounced antiseptic properties, the herb can facilitate healing of scalp conditions as well as returning the sheen to dull, lifeless hair. It is also widely employed as a diuretic.

Garlic's high profile today is reflected by its inclusion in Deborah C. Harding's *The Green Guide to Herb Gardening: Featuring the 10 Most Popular Herbs* (2000). Discount Natural Herbs, a leading Internet retailer, listed garlic in its "Top Ten Selling Herbs" in mid-2001. Celebrities such as radio and television interview host Larry King (Garlique ads) were highly visible in the 1990s touting the preventative and curative role of garlic in heart disease. These claims are based on the findings of more than twenty scientific studies that garlic reduces overall and harmful LDL cholesterol levels, serum triglycerides, and blood pressure, in addition to inhibiting cholesterol oxidation and platelet aggregation (the tendency of the blood platelets to clump).

The findings of a four-year study featuring 152 men and women (randomly assigned to take either a placebo or 900 milligrams of standardized garlic powder daily)—the longest clinical trial to evaluate the effects of a dietary supplement on reducing heart attack risk to that point—documented that a garlic powder supplement can help prevent and, in some cases, even reverse plaque buildup in the arteries (McCaleb, 2000, Web site). Participants all possessed advanced plaque accumulation (long associated with an increased risk of heart attack and stroke), as well as at least one other established risk factor for heart disease (e.g., high cholesterol, high blood pressure, diabetes, or a history of smoking). The study found that those participants taking garlic had a 2.6 percent reduction in plaque volume, whereas the placebo group experienced a 15.6 percent increase. The findings supported the scientific case for garlic as a "pleiotropic" substance, meaning that the herb's mild effects on many different measurements of heart health add up to significant overall benefits.

Perhaps the greatest drawback to the use of garlic is "annoyance by odor." Furthermore, those taking the herb in natural form have reported occasional gastrointestinal discomfort (e.g., heartburn).

BIBLIOGRAPHY

Harding, Deborah C. *The Green Giude to Herb Gardening: Featuring the 10 Most Popular Herbs.* St. Paul, MN: Llewellyn, 2000.

Heinerman, John. *From Pharaohs to Pharmacists: The Benefits of Garlic.* New Caraway, CT: Keats, 1994.

Heinerman, John. *Heinerman's New Encyclopedia of Fruits and Vegetables.* Paramus, NJ: Prentice-Hall, 1995.

Hollis, Sarah. "Garlic." In: *The Country Diary Herbal.* New York: Henry Holt, 1990, pp. 57-59.

Koscielny, J., Klossendork, D., Latza, R., Schmitt, R., Radtke, H., Siegel, G., and Kiesewetter, H. "The Antiatherosclerotic Effect of Allium Sativum." *Atherosclerosis 144* (1999): 21.

McCaleb, Rob. "Studies Question Garlic Effectiveness." *Herb World News Online.* Boulder, CO: Herb Research Foundation, 2000. 2p. www.herbs.org/current.

Morien, Krista. "Standardized Garlic Powder Reduces Heart Attack Risk." *HerbalGram 47* (1999): 21. www.herbalgram.com.

GINGER; WILD GINGER

Wild ginger *(Asarum canadense)*, a member of the birthwort family, is native to the eastern woodlands of North America. (In addition, a wide variety of wild gingers—approximately 100 species in all—are native to the tropics and subtropics of eastern Asia.) It is a hardy plant capable of surviving temperatures as low as forty degrees (Fahrenheit) below zero.

The growth cycle begins in early spring when wild ginger rhizomes on ground level or immediately below the surface develop a new pair of soft, hairy, kidney-shaped leaves. By fall, these leaves will have matured to as much as seven or eight inches in width atop sprawling stalks approximately fifteen inches in length. A reddish-brown flower with no petals appears on top of the ground between the two leaf stalks by May. Believed to be self-pollinated, the bell-shaped portion of the flower—the calyx, one-half inch long and one and one-half inches across, with three pointed lobes—remains until large, oval, glossy grayish or brown seeds ripen in four to six weeks.

Medicinal uses of wild ginger extend back to before European colonization of the New World. Decoctions and infusions of the wild ginger rhizome were employed by Native Americans to induce menstruation and regulate an irregular heartbeat. The Meskwaki were known to have steeped crushed rhizomes and poured the liquid into the ear to relieve an earache.

The European colonists appear to have appropriated various preparations incorporating the herb from Native Americans, most notably a tooth powder that blended powdered wild ginger rhizome with the pulverized bark of black alder, bayberry, and black oak. These settlers also adapted the plant to recipes employing the unrelated gingerroot *(Zingiber officinale)*, a staple Old Country herb. This influence was found in the use of the candied rhizome and syrup to relieve flatulence and stomach cramps as well as the "Doctrine of Signatures," the notion held by many cultures that plants have been signed by their Creator with visible clues to their usefulness. In this case, the shape of the wild ginger leaves was linked with the treatment of kidney disorders. Additional colonial-era folk remedies involving wild ginger included treating snakebite (which gave rise to one of the popular names for the plant, Canada snakeroot) and inducing sweating to relieve fever.

In recent years, studies in Denmark, Italy, and elsewhere have revealed that ginger is as effective as Dramamine (dimenhydrinate), or more so, in the prevention of motion sickness. Furthermore, ginger users have been found to experience fewer side effects, notably sleepiness.

European settlers and Native Americans alike utilized wild ginger to flavor foods. The former dried and grated the root as a substitute for gingerroot. The plant has also found increasing favor in shade or woodland gardens, whether massed as an edging, ground cover, or large clump or interplanted with other shade lovers.

Gingerroot's common name in most languages, and generic name, *Zingiber,* represents a derivation of the Sanskrit word *shringavera,* meaning "shaped like a deer's antler" (a reference to the knobby shape of the rhizome). The spice possesses a long and rich history. During the fifteenth century, gingerbread was given as a symbol of love and respect. The practice of sprinkling ginger on top of beer or ale, and then stirring it with a hot poker, resulted in the invention of ginger ale. Most ginger is now grown commercially in Jamaica, followed by India, Africa, and China; a relatively small amount has been cultivated in Florida.

In addition to its widespread use in Chinese, Japanese, and Ayurvedic medicine (the latter of which was popularized stateside by pop physicians such as Deepak Chopra in the 1990s), gingerroot has long been a popular flavoring in Far Eastern, Indian, Indonesian, Sri Lankan, Himalayan, African, and European cuisines. Today, its appeal appears to be just as great in North America. McCormick's *Flavor Trends 2000* included the spice in its "Top 10 Flavors for 2000."

BIBLIOGRAPHY

Albert-Matesz, Rachel. "The Joys of Ginger." *The Herb Companion 13*(1) (October/November 2000): 20-23.

Heinerman, John. "Ginger *(Zingiber officinale).*" In: *Heinerman's Encyclopedia of Healing Herbs & Spices*. Paramus, NJ: Prentice-Hall, 1996, pp. 251-155.

"Herbal Healthwatch: Ginger to the Rescue for Motion Sickness." *The Herb Quarterly* (Winter 1999): 7.

Strauch, Betsy. "An Herb To Know." *The Herb Companion 8* (December/January 1995/1996): 32-33.

GINKGO

Ginkgo biloba is a slow-growing tree with heavy, gangly-looking limbs capable of reaching forty meters in height. Its leaves, fan-shaped with two lobes, possess a deep green color, and are thick and tough in texture. The species includes separate male and female trees: the former produce pollen, while the latter bear fruit that are close to apricots in appearance. The fruit combines the smells of rancid butter and vomit; furthermore, a rash can result from handling it.

The ginkgo tree originated during the Jurassic period, spreading across the cold northern forest belt. Early mammals ate the tree's nut-like seeds and helped to disperse them; forests comprised almost entirely of the ginkgo arose. Individual trees are extremely long-lived, lasting two thousand years or more. Their leaves, branches, and roots have been found to contain a powerful antioxidant and healing chemicals that aid in defying the aging process.

The ice ages almost eradicated the ginkgo tree; its long maturation period hindered the spread of seeds to warmer climes to the south. However, the shamans of ancient China, considering the tree a valuable source of food and medicine, actively cultivated it. Later, Buddhist monks would plant it in their monastery gardens. The *Shennong Bencaojing,* credited to the legendary Emperor Shen Nong (3494 B.C.) and considered by many to be the first book on herbalism, discussed the use of ginkgo leaves for boosting memory and treating breathing problems such as asthma and bronchitis.

According to Jonathan Zuess, Chinese medical texts continued to find new uses for ginkgo throughout history:

> The leaves were used externally, as a poultice to help heal wounds and to treat sun damage (freckles), and internally, to improve circulation to the limbs in cases of frostbite. The fruit pulp was used as a treatment for infections and intestinal parasites. . . . In traditional Chinese medicine ginkgo seeds are classified as a "kidney yang tonic," meaning they can increase energy in the

> urinary and sexual organs. They're used to treat impotence, bedwetting, and bladder problems. They're also good for treating disorders of the ear, like deafness. (American Botanical Council Web site)

A German surgeon, botanist, and adventurer, Engelbert Kaempfer, was captivated by the ginkgo tree and brought seeds back to his homeland around 1730. Given its hardiness against pollution and pests, it rapidly came to be recognized as an ideal ornamental tree in the post–Industrial Revolution urban centers in Europe. From there, the first ginkgo tree was imported to the United States in 1784.

It was not until the 1960s, however, that ginkgo gained widespread acceptance among Western physicians. In contrast to British and U.S. medical professionals, European doctors had long been incorporating herbs into their treatment regimens. By the 1990s, ginkgo had become the most frequently prescribed herb in Europe. Faced with considerable scientific evidence from abroad that herbal medications could be an effective form of treatment, American doctors began using ginkgo in quantities. W. Schmid noted,

> To keep up with the growing demand, herbal companies have established large plantations of ginkgo trees on the Atlantic coast of France, in South Carolina, and in China. Ginkgo is also grown commercially in Japan and South Korea. (p. 755)

Ginkgo is commercially available in a variety of configurations; these are sold both through online distributors and bricks-and-mortar health food stores. All Western preparations utilize the leaves, which are harvested in fall (when the concentration of active ingredients is highest) and then finely ground and soaked in a solution of acetone and water. After the active ingredients have dissolved into the liquid, it is strained to remove solid residues such as cellulose. Following additional purification techniques aimed at adjusting the concentrations of the major active ingredients, the extracting liquid is largely removed, resulting in a gooey, greenish-brown syrup. Ginkgo leaf powder is then added as a binder, and the substance is packaged in capsules or gel tabs.

Ginkgo can also be obtained in the following forms, which feature minimal processing:

- The leaves are freeze-dried, ground up, and packaged into capsules
- Ground leaves are soaked in alcohol or glycerine, strained, and the unrefined liquid extract marketed in small brown glass bottles
- Fresh or dried ginkgo leaves are brewed as tea

Perhaps the most notable feature of ginkgo's present-day success in North America is its ability to cut across demographic lines. It is in high demand as a means of increasing oxygen to the brain—thereby enhancing mental sharpness—among yuppies, latter-day hippies, hardcore advocates of alternative medicine, second- and third-generation Asian Americans, and many other subcultures. Its inclusion in Discount Natural Herbs' "Top Ten Selling Herbs" as of mid-2001 is further evidence of the herb's popularity.

BIBLIOGRAPHY

American Botanical Council. *German Federal Commission E Monograph: Ginkgo.* 1994. Available at: www.herbalgram.org.

Halpern, G. *Ginkgo: A Practical Guide*. New York: Avery, 1998.

Kleijnen, Jos., and Knipschild, Paul. "Ginkgo Biloba." *Lancet 340* (1992): 1136-1139.

Le Bars, Pierre L., and Katz, Martin M. "A Placebo-Controlled, Double-Blind, Randomized Trial of an Extract of Ginkgo Biloba for Dementia." *Journal of the American Medical Association 267*(16) (1997): 1327-1332.

Schmid, W. "Ginkgo Thrives." *Nature 386* (1997): 755.

Zuess, Jonathan. *Ginkgo: The Smart Herb*. New York: Three Rivers Press, 1998.

GINSENG

Ginseng (referring to twenty-two plants largely within the Araliaceae family, the best known of which is probably *Panax quinquefolius*) is one of the most popular herbs among health food enthusiasts and users of medicinal plants. The plant has been used by the Chinese for thousands of years. In ancient times, the root was thought to resemble a human being. In a manifestation of the "Doctrine of Signatures," early Chinese healers concluded that ginseng could be employed to help man.

Ginseng is most widely used today as a mild stimulant and relaxant. However, it has also enjoyed considerable popularity for its endurance capabilities and as a means of enhancing sexual potency. People in many cultures worldwide consider the herb to be the ultimate aphrodisiac. Some credence was given to this belief when Japanese scientists identified sex hormone activity in ginseng preparations administered to male and female rodents alike.

Despite this notoriety, GNC, a leading health food chain, touts its Herbal Plus Panax Ginseng (Full Spectrum and Standardized versions) in a rather ambiguous manner, simply noting that their premium line of herbs "take consistency, purity, potency and maximum bioavailability to the next level." Unfortunately, many products unrelated to ginseng either botanically or chemically are often passed off fraudulently as the original root.

BIBLIOGRAPHY

Garland, Sarah. "American Ginseng." In: *The Complete Book of Herbs & Spices*. Pleasantville, NY: Reader's Digest Association, 1993, pp. 87, 269.

Griggs, Barbara. *Green Pharmacy: The History and Evolution of Western Herbal Medicine*. Rochester, VT: Healing Arts, 1997.

Heinerman, John. "The Ginsengs." In: *Heinerman's Encyclopedia of Healing Herbs & Spices*. Paramus, NJ: Prentice-Hall, 1996, pp. 257-259.

McHoy, Peter, and Westland, Pamela. *The Herb Bible*. New York: Barnes & Noble Books, 1994.

GOLDENROD

Delaware, perhaps best known as the first of the original thirteen colonies to ratify the United States Constitution, achieved further renown when it became the first state to adopt an herb as part of its official regalia. Goldenrod, already singled out as the official flower of Kentucky and Nebraska, achieved this status on June 24, 1996.

Goldenrod—an indigenous plant popular in Delaware and elsewhere as an herb tea, a valued element of floral arrangements, a source of dye, and simply an accent in the wilds—has been used by European settlers and Native Americans alike at least as far back as the early seventeenth century. Goldenrod's most notable role in American history probably resulted from the Boston Tea Party, whereby rebellious colonists launched a boycott of English tea by depositing a load of it into the harbor to protest outrageous tariffs. The Americans went on to drink Liberty Tea, an herbal concoction made from goldenrod and other native plants.

In the 1890s, some Delaware citizens pushed unsuccessfully to have goldenrod named the state flower. Despite losing out to the peach blossom on May 9, 1895, the graceful yellow blossoms of the sweet goldenrod have remained a familiar sight at the sandy edges of woodlands and roadsides throughout Delaware and eastern North America. During the early 1990s, the Delaware Herb Growers and Marketers Association—actively involved in the growing of such plants for restaurants and farmers' markets, the sale of herbal products, the supply of material for the dried herb and essential oil markets, and herb research and education—lobbied successfully to name goldenrod the state herb. Association members noted that goldenrod was not a narcotic or hallucinogen and wouldn't be a political embarrassment at a later date. It was also pointed out that the plant has been unjustly cited as a problem for allergy sufferers; rather, it is insect-pollinated, and most species (unlike ragweed) disperse minimal pollen in the wind. Most important, the members argued that having a state herb would be a symbol of support for statewide growers in addition to increasing an awareness of herbs among Delaware citizens.

BIBLIOGRAPHY

Garland, Sarah. *The Complete Book of Herbs & Spices*. Pleasantville, NY: Reader's Digest Association, 1993.

Tucker, Arthur O. "The Official Goldenrod." *The Herb Companion* (February/March 1997): 24-25.

GOLDENSEAL

A small perennial also known as yellow root, goldenseal *(Hydrastis canadensis)* grows wild in woodland areas of the eastern United States. The thick, knotty, yellow rootstock nurtures a hairy stem; near its top, roughly a foot high, grow a pair of five-lobed, serrated leaves. A single greenish-white flower appears in late spring; the sepals fall away as the flower opens, leaving a mass of stamens.

The root was very popular among Native Americans as a medicine. Cherokees also employed it to dye their skin and clothing bright yellow. European settlers would later adopt it for similar purposes. Modern-day physicians, alternative-care doctors, and holistic healers continue to find a wide range of uses for the plant.

The March 1983 issue of the *Indian Journal of Ophthalmology* (cited in Heinerman, 1996, p. 261) reported that the alkaloid berberine (which is responsible for the root's yellow color) is highly effective in treating inflammations of the cornea and iris caused by herpes. Doctors also use goldenseal root to cure many other eye ailments, particularly conjunctivitis.

Additional medicinal applications of goldenseal include:

- Reducing yeast infection
- Healing sores in the mouth or on the gums and tongue
- Cleansing the system as part of a drug withdrawal program
- Easing discomfort caused by (and healing) poison ivy rash
- Reducing insulin dependency
- Relieving sinus problems
- Inhibiting blood clotting

A number of caveats should be noted in utilizing the herb. Patients prone to strokes are advised not to ingest goldenseal capsules if they have hypoglycemia. If taken for longer than two months, it can compromise the colon's ability to manufacture B vitamins. Likewise, pregnant women using it risk undesirable side effects.

BIBLIOGRAPHY

Garland, Sarah. *The Complete Book of Herbs & Spices*. Pleasantville, NY: Reader's Digest Association, 1993.

Griggs, Barbara. *Green Pharmacy: The History and Evolution of Western Herbal Medicine*. Rochester, VT: Healing Arts, 1997.

Heinerman, John. "Goldenseal *(Hydrastis canadensis)*." In: *Heinerman's Encyclopedia of Healing Herbs & Spices*. Paramus, NJ: Prentice-Hall, 1996, pp. 261-266.

H

HENNA

A perennial shrub sometimes reaching twenty feet in height, henna *(Lawsonia inermis)* possesses narrow, gray-green leaves and small, fragrant red or cream flowers followed by blue-black berries. Its physical attractiveness has led to it being employed as an ornamental hedge. Native to Upper Africa, Asia, and parts of Australia, it has been naturalized and cultivated in many locales worldwide, most notably the American tropics, Egypt, India, and the Middle East.

In dried, powdered form, the leaves are cooling and astringent. Mixed with water to form a paste, they can be applied to the forehead for headaches and fevers as well as elsewhere for swelling, bites, or stings.

Henna is best known as a cosmetic dye. Since antiquity, the dried, ground leaves (which contain the active ingredient lawsone) have been soaked in water and lime juice in the manufacture of orange, red, and brown dyes renowned for holding their color. Egyptian women have long utilized it to stain the palms of their hands and soles of their feet to prevent perspiration and blistering from exposure to the sun. Mummies dating back several thousand years retain some of the henna dye used to decorate their nails. Indian and Pakistani men and women also colored their nails, fingers, hands, and hair with the dye.

During the 1990s, henna achieved renewed popularity in the West as both a natural hair coloring and a major ingredient in body painting. Temporary tattoos employing henna dye last about a month; they represent a healthful alternative to the traditional inks of permanent tattoos.

John Heinerman notes that the Zabbaline, Cairo's Coptic Christian minority known to engage in urban scavenging for a livelihood, use henna to combat skin problems arising from their daily activities:

> Considered objects of scorn by most Cairo inhabitants, the zabbaline suffer tremendous exposure to disease. As I walked among their homes and yards filled to excess with huge piles of rotting garbage, I couldn't help but notice how many of the children had open sores on their arms, hands, and legs. But upon closer examination of some of them, I was amazed to find *no* apparent spread of infection. Instead, such sores were covered with a rust-colored paste, which I later came to understand was henna powder mixed with water and applied over the sores to keep them from becoming worse. Since then I've recommended henna paste to those suffering from herpetic lesions and the sores common to AIDS patients with pretty good results in their healing. (p. 275)

BIBLIOGRAPHY

Garland, Sarah. *The Complete Book of Herbs & Spices*. Pleasantville, NY: Reader's Digest Association, 1993.

Heinerman, John. "Henna *(Lawsonia inermis)*." In: *Heinerman's Encyclopedia of Healing Herbs & Spices*. Paramus, NJ: Prentice-Hall, 1996, pp. 273-274.

HERB CRAFTS

Prior to the Industrial Revolution, herbs occupied a central place in domestic crafts. They remain popular up to the present day for dyeing and decorative activities.

The most prevalent decorative uses of herbs include dried floral arrangements and wreaths. Many of these evolved out of symbolic or religious beliefs. Notable examples include hanging mistletoe to bring good fortune and fertility, and tying St.-John's-wort in bunches at a window to ward off evil spirits. Herbs in dried arrangements offer a colorful alternative to the typical muted brown and green shades. The box provides a listing of herbs that retain a rich color when dried.

Herbs Preferred for Dried Floral Arrangements and Wreaths

Bay (leaves and stems)	deep green color
Bee balm (flowering stems)	scarlet
Caraway (seed heads)	tan; brown
Chamomile (flowers)	yellow-gray
Chervil (seed heads)	green; brown
Dill (seed heads)	green; brown
Eucalyptus (leaves)	silvery gray
Fennel (seed heads)	light gray-brown
Lady's Mantle (flowering stems)	yellow-green
Lavender (flowering stems)	purple; dark blue
Pennyroyal (flowering stems and leaves)	purple
Rosemary (stems and leaves)	deep green
Sorrel (seed heads)	shades of green
Southernwood (leaves and stems)	green
Tansy (flowering stems)	bright yellow
Yarrow (flowering stems)	pink; white; deep yellow

Source: Garland, 1993.

While most seasonal decorations today that use herbs are regarded as purely ornamental, symbolic meanings of herbs have been preserved in Victorian "language of flowers" listings. Such information is still useful for herbal arrangements geared to special occasions.

Symbolic Meanings of Selected Herbs

Basil	Best wishes; warm friendship
Bay	Honor; loyalty; unchanging affections; faithfulness
Borage	Courage
Chamomile	Patience; meekness; resignation
Chervil	Sincerity; warm feelings
Coriander	Hidden worth; concealed feelings
Elderflower	Compassion; sympathy
Evening primrose	Inconstancy; uncertainty
Fennel	Strength; flattery
Geranium	Comfort; consolation
Lavender	Silence; recognition/acceptance of love
Marjoram	Happiness; joy

(continued)

(continued)

Mint	Wisdom
Myrtle	Love; first declaration of love
Nasturium	Conquest
Parsley	Celebration; festivity
Red rose	True love
Rose	Silence
Rosemary	Remembrance
Rue	Repentance
Saffron	Marriage
Sage	Esteem; friendship
Southernwood	Constancy
Thyme	Activity

Although synthetic dyes dominate the modern-day marketplace, many craft weavers and spinners continue to use herb-based dyes for the subtle and variable shades they provide. Herbs widely employed for dyeing include elder (leaves, berries, bark), marigold (petals), parsley (leaves, stems), St.-John's-wort (flowers), Tansy (flowering tops), and woad (leaves).

BIBLIOGRAPHY

Garland, Sarah. "Household Herbs and Spices." In: *The Complete Book of Herbs & Spices*. Pleasantville, NY: Reader's Digest Association, 1993, pp. 220-243.

Hollis, Sarah. "Home-Made Herbal Products." In: *The Country Diary Herbal*. New York: Henry Holt, 1990, pp. 110-113.

McHoy, Peter, and Westland, Pamela. "Crafts"; "Wreaths & Garlands." In: *The Herb Bible*. New York: Barnes & Noble Books, 1994, pp. 196-197; 198-199.

HERB FAIRS

Herb fairs represent a spin-off of a longtime American tradition. Former U.S. president William McKinley noted, "Expositions are timekeepers of progress" that stimulate the "energy, enterprise, and intellect of the people and quicken human genius" (Fisher, 2001, p. 31). Whether referred to as expositions or some other appellation, fairs predate the founding of this country as a device for showcasing innovations in the arts and technology, stimulating business and trade, and facilitating local pride.

The format for these events—most notably, diversified entertainment, educational opportunities, the display and sale of goods, a centralized location often organized along the lines of a small-scale city, and an extended time frame generally ranging from several days to longer than a year, all structured around a central theme or purpose—has remained essentially unchanged since the advent of the groundbreaking fairs in medieval Europe. The rise of events focusing on the beneficial aspects of botanicals corresponds with the herb renaissance of the post-1960s era. By the 1990s, a wide range of special events surrounding this phenomenon—sometimes touting a specific herb or spice—were open to the public at large.

The majority of these events are promoted via fan organizations, specialty publications, and the Internet. As a case in point, a browser search of the Net revealed that the following herb fairs were active through 2000:

- *American Herbalist Guild's Annual Herb Symposium.* The event is held every October at Mount Madonna, California. <www.AmericanHerbalistsGuild.com>; e-mail: ahgoffice@earthlink.net.
- *The Days of Wine and Lavender.* Matanzas Creek Winery, Santa Rosa, CA. Pricey one-day affair incorporates new wine releases, lavender cuisine, live music, dancing, lavender games, and photography portraits.
- *DeBaggio's Lavender Festival.* DeBaggio Herbs, Chantilly, Virginia. Includes lectures, demonstrations, lavender crafts, cookings, and guided tours of lavender and herb gardens. <www.members.aol.com/tdherbs>.
- *Family Ozark Herbal Odyssey.* Heritage Herb Gardens, Mountain View, Arkansas. Week-long seminar featuring leading herbalists.
- *Frontier's Annual HerbFest.* Frontier Natural Products Co-op, Norway, Iowa. Includes herbal education programs, health issues, and family entertainment. <http://www.frontiercoop.com/herbfest/98/index.html>.
- *Gilroy Garlic Festival.* <http://www.gilroygarlicfestival.com/>.
- *Herbfest.* Herb Society of Manitoba. The third annual installment, entitled "Herbs for the New Millennium," was held July 8, 2000, at the Assiniboine Park Conservatory and grounds, Winnipeg.

- *Herb Society of America's Education Conference and Annual Meeting.*
- *Herbal Delights.* Callaway Gardens' Education Department. Workshops held September 9, 2000, covered creating a winter container garden, growing herbs in the shade, and learning to appreciate and use herbal fragrances. E-mail: education@callawaygardens.com.
- *Hudson Valley Garlic Festival.* Kiwanis Club of Saugerties, Saugerties, New York. Includes performing arts, food vendors, arts and crafts vendors, and education and instruction. <http://www.hopefarm.com/garlic.htm>.
- *International Herb Association Annual Conference.* Saskatchewan Herb and Spice Association, Saskatoon, Canada. Program includes presentations and workshops on culinary herbs, propagation, medicinal tracts, and marketing and business.
- *Kord Farm Lavender Festival.* Focuses on the growing, care, and uses (e.g., health, crafts) of lavender.
- *Pennsylvania Herb Festival.* Introduction to herbs; participants are encouraged to touch and smell the plants. Covers the varied uses of herbs, including aromatherapy, cooking, cleaning, cosmetics, healing, decorating, and as jewelry. Includes guest speakers, book signings, and business exhibits.
- *Sequim Lavender Festival.* Annual showcase held in Sequim, Washington along with the Lavender Growers Conference. <www.lavenderfarms.com>.

BIBLIOGRAPHY

Findling, John E., and Pelle, Kimberly D. (Eds.). *Historical Dictionary of World's Fairs and Expositions.* Westport, CT: Greenwood, 1990.

Fisher, Jack C. *Stolen Glory: The McKinley Assassination,* La Jolla, CA: Alamar, 2001, p. 31.

Manning, Martin. "Fairs! Fairs! Fairs! The United States Information Agency and U.S. Participation at World's Fairs Since World War II." *Popular Culture in Libraries* 2(3) (1994): 1-32.

Rydell, Robert (Ed.). *Books of the Fairs.* Chicago: American Library Association, 1992.

HERB GARDENS

During antiquity, gardens attached to temples or sacred groves were used for the cultivation of medicinal plants. Sarah Garland notes that the basic plan of these gardens was determined by the necessity of growing and tending plants within a confined space as well as protecting them and providing irrigation. Accordingly, herbs and other botanicals were laid out in straight lines punctuated by paths and systems of water channels or a pool, all of which was generally surrounded by a wall of some kind. She adds, "These practical considerations combined to create the formal, rectilinear garden design familiar in ancient Egyptian frescoes, early Persian and Islamic paintings, and in the Roman *hortus* with its tidy divisions between the vegetables and herbs, and colonnades and courtyards filled with terracotta pots of aromatic plants" (p. 120).

While many of the cultivated gardens of Imperial Rome reverted to wilderness during the Middle Ages, the few that remained in southern Europe provided the pattern for the earliest monastery gardens. There, the traditions of herb culture were refined by clerics in accordance with a lifestyle emphasizing self-sufficiency, an affinity for nature, and administering to the needs of the less fortunate. Through herb gardens, the monks were able to exchange cuttings and seeds, record the properties of plants, and create and disseminate manuscript herbals. By the Renaissance, the utilitarian herb garden had become increasingly popular and elaborate. The beds formed looping patterns and mazes, and were often bordered with clipped hedges of lavender or germander and pebbles or bones, and arranged in the fashionable Italianate style.

The post-Renaissance era saw the rise of physic gardens supplying medicinal herbs for apothecaries and clerics; travelers and explorers in the sixteenth and seventeenth centuries provided new species (e.g., the nasturtium, the sunflower). While established models were being carried to the New World, large botanical gardens arose in Europe, attached to universities and medical schools in order to facilitate botanical and health care study.

Inventory of Chief Characteristics of Formal and Informal Herb Gardens

Formal Garden Plan

- Design generally follows a clear pattern with paths and beds laid out to balance and complement each other.
- Geometric shapes—curves, rectangles, triangles, and circles—are often the basis of a particular plan.
- The focal point is usually the center of the garden, marked by a sundial, a seat, a large pot or statue, or visually arresting plant (e.g., ornamental tree).
- Plantings tend to be sparse with herbs spaced well apart, often in straight lines or groups; low walls or wooden planks often serve to separate species.
- A well-maintained low hedge of perennial herbs is frequently used as a traditional edging.
- Communicates a sense of order and peace.

Informal Garden Plan

- Gives an impression of barely controlled abundance, of unexpected forms arising from the natural configuration of the local landscape.
- Herbs are typically grown close together; species that will not obliterate one another should be selected.
- Color and texture communicate an impressionistic, casual effect.
- Clipping and shaping should be kept to a minimum.
- Walls, paths, and other man-made features should be composed of stone, brick, or other local material.

Source: Garland, S., 1993, p. 175.

The Romantic movement beginning in the late eighteenth century gave rise to a less formal conceptualization of herb gardens. The studied carelessness and natural, asymmetrical composition of gardens then in fashion has remained an influence even up to the latest resurgence of interest in the cultivation and use of herbs that began in the 1960s.

By the 1980s, gardening had become a national obsession in the United States. Through the 1970s, the archetypal gardener was likely to be over fifty with time and money to spare. By 1988, a cover story in *Time* magazine noted that 78 percent of American households had acquired the gardening habit; recent surveys indicated that the most fervent converts were thirty to forty-nine years old. Baby boomers accounted for half of the (then) record $17.5 billion spent on horticultural products in 1987. While much of this money was spent on vegetable products and other mainstream fare, Nancy Gibbs reported that herbalists had become major players within the field:

> Last year 6 million households spent $46 million growing herbs, in contrast with 5 million spending $39 million the year before. Some make tea from them, some bathe in them, some swear to their healing powers. Having mastered the basic basil, rosemary and sage, gardeners move on to lovage stems, beebalm blossoms and lemon grass. The health conscious prize herbs as a salt substitute, while the cost conscious find that pricey, herb-flavored vinegars and oils are easily made at home. (p. 62)

By the 1990s, herb gardening had become increasingly more visible on the American landscape. Many local communities were graced by educational programs and beautification projects sponsored by herb appreciation societies. Furthermore, the individual enthusiast had a wealth of information and supplies to choose from, ranging from traditional print publications to Web-based retailers.

BIBLIOGRAPHY

Bacon, Richard M. *The Forgotten Arts: Growing, Gardening, and Cooking with Herbs*. Dublin, NH: Yankee Publishing Inc., 1972.

Beckett, Kenneth A. *The Garden Library: Herbs*. New York: Ballantine Books, 1984.

The Beginner's Herb Garden. Mentor, Ohio: Herb Society of America, 1983.

Bown, Deni. "Designing an Herb Garden." In: *Encyclopedia of Herbs and Their Uses*. New York: Dorling Kindersley, 1995, pp. 20-41.

Coy, Cissie. "The Herb Gardens at Washington Cathedral." *The Herb Companion* (December/January 1990/1991): 41-45.

Davies, Jill. *Herbs and Herb Gardens*. Aylesbury, U.K.: Shire Publications, 1983.

Garland, Sarah. "The Herb Garden." In: *The Complete Book of Herbs & Spices*. Pleasantville, NY: Reader's Digest Association, 1993, pp. 120-157.

Gibbs, Nancy R., with Mary Cronin, James Willwerth, and O. Freidrich. "Paradise Found: America Returns to the Garden." *Time 131* (June 20, 1988): 62-71.

Halloran, Kathleen. "A National Treasure." *The Herb Companion* (October/November 1993): 42-48.

Hobhouse, Penelope. *Gardening Through the Ages*. New York: Simon & Schuster, 1992.

Hopkinson, Simon, and Hopkinson, Judith. *Herbs*. Chester, CT: Globe Pequot Press, (Includes chapters entitled "Herb Gardens," "Herb Garden Maintenance," "A Selection of Garden Herbs," and "Gardens to Visit"), 1989.

Paterson, Allen. *Herbs in the Garden.* London: J.M. Dent & Sons Ltd., 1985.
Reid, Shirley. *Herbs for the Home and Garden.* Topsfield, MA: Salem House, 1988.
Sanecki, Kay N. *History of the English Herb Garden.* Foreword by Anthony Huxley. London: Ward Lock, 1992.

HERB GROWING & MARKETING NETWORK

Formed in 1990 and currently more than 2,000 members strong, the Herb Growing & Marketing Network is the largest trade association in a rapidly growing industry. It is geared to serving three distinct groups:

- Anyone interested in establishing some kind of herb business, whether its growing herbs commercially, running an herb shop, manufacturing herbal products, or maintaining a healing practice
- Small- to medium-sized businesses attempting to expand sales
- Large firms desiring more exposure at minimal cost

The organization is, first and foremost, an information service. It maintains a core library of more than 3,000 books and subscribes to all relevant periodicals. It monitors relevant Internet mailing lists and searches the World Wide Web for research resources of potential value to members. Publications include an industry-wide annual resource guide, the *Herbal Green Pages;* a quarterly trade journal, *The Bu$iness of Herbs;* national conference proceedings; and various booklets reflecting the needs and interests of the membership.

BIBLIOGRAPHY

Herbworld Online Home Page: www.herbalconnection.com. Copyright 2001. 2p. Includes the following links: "Join Now," "Green Pages," "Crop Shop," "Speakers Bureau," "Herbalpedia," "Herbal Exchange," "Business Startup," "Recipes," "Links To Friends," and "Contact Us."

HERB JARGON

Similar to many subjects or fields of endeavor, herb research and practice has spawned its own subculture. This subculture includes a system of beliefs, code of ethics, and specialized language. In addition to biological terms not familiar to the general public, the language of herb fanciers is typically peppered with words that describe the varied nutritional, medicinal, cosmetic, aesthetic, and recreational uses of these plants. The glossary below reflects some of the discipline's more germane terms.

antiseptic: fights infection by destroying or inhibiting the growth of bacteria.

astringent: draws together and contracts organic tissues.

decoction: an extract obtained by boiling or simmering tough portion of a plant in some type of liquid in order to draw out its active principles.

demulcent: a substance that soothes or relieves irritation.

distillation: the process of separating the volatile from the fixed portion of a substance through evaporation and condensation.

emetic: an agent that cases vomiting.

expectorant: an agent that serves to purge the bronchial tubes of phlegm.

herbal: a guidebook that aids in the identification, cultivation, and use of herbs.

infusion: a means of extracting the active principles of an herb by steeping it in water.

lanceolate: long, narrow, tapering, spear-shaped plant leaves.

mordant: A substance such as wood ashes, salt, vinegar, soda, urine, or yogurt used to fix herb dyes, to ensure that they are colorfast and thoroughly absorbed by the fiber.

palmate: a leaf possessing more than three leaflets radiating from a basal point; comparable to fingers projecting from the palm of a hand.

panicle: grouped flowers.

petiole: leaf stalk.

pinnate: a leaf split into four or more leaflets arranged in two rows along a stalk.

potpourri: dry or moist aromatic mixtures of flowers, herbs, and spices kept in covered pots or bowls; warmed and opened to release their scent. Extremely popular in Europe during the seventeenth and eighteenth centuries.

principle: a plant's active chemical constituent.

rhizome: creeping, typically swollen, underground stem.

sepals: outer flower parts combined to form the calyx.

stamen: a plant's male reproductive organ.

stigma: the part of the plant style receiving pollen during fertilization.

strewing: distributing or hanging herbs in buildings in order to repel insects, overpower musty and unsanitary smells, reduce the temperature (oxidizing oils form an aromatic ozone), etc. Very popular in Britain and northern Europe from the Middle Ages until the late eighteenth century.

style: extension of a plant's ovary.

tisane: an infusion or tea prepared with one or more herbs and spices.

trifoliate: leaf possessing three leaflets.

tuber: Swollen part of an underground root or stem.

umbel: a flower head possessing stalks that radiate from one point.

BIBLIOGRAPHY

Garland, Sarah. *The Complete Book of Herbs & Spices*. Pleasantville, NY: Reader's Digest Association, 1993.

McHoy, Peter, and Westland, Pamela. *The Herb Bible*. New York: Barnes & Noble Books, 1994.

HERB QUEST INTERNATIONAL

The herbal renaissance has stimulated the development of a broad-based infrastructure concerned with the acquisition and dissemination of information. Herb Quest International, an Internet search engine created in 1998, caters to business professionals concerned with the manufacture, distribution, and retailing of herbs.

Herb Quest International also provides members with links and ads as well as online boards and discounted publications. One such publication is *Herbalpedia,* a CD-ROM utilizing Adobe Acrobat software that includes 257 monographs covering botanicals, essential oil plants, edible flowers, and spices. Including more than 800 pages of information, its aim is to increase the user's knowledge in addition to augmenting culinary and healing skills. To assist CD-ROM users, Adobe Acrobat Reader (Version 4 is preferred) can be downloaded for free at www.adobe.com.

BIBLIOGRAPHY

"Herb Quest Search Engine & Herbal Information Provider," www.herbquest.com. Herb Quest International. (March 2001) 2p.

HERB RESEARCH FOUNDATION

The Herb Research Foundation (HRF) is considered the world's most comprehensive source of accurate scientific information on medicinal plants. It is a nonprofit research and education organization dedicated to improving world health through the informed use of herbs. Since its founding in 1983, the HRF has been educating the public, health practitioners, legislators, and the media about the health benefits and safety of herbs, drawing on a specialty botanical library containing more than 300,000 scientific papers on thousands of herbs and multiple online sources of clinical, pharmacological, chemical, and horticultural information. The HRF was the brainchild of Robert S. McCaleb, an internationally recognized authority on scientific and regulatory issues affecting herbs, and Mark Blumenthal, owner of an Austin-based operation called Sweethardt Herbs. Appalled by the low standards of the industry, he played a key role in founding the Herb Trade Association in the late 1970s. After the HTA died in 1981 from a general lack of interest, Blumenthal helped establish both the American Herbal Products Association and the HRF.

The HRF was created in response to the need for a bridge between modern American medical practice and growing public interest in the use of botanicals for health and medicine. Today, through media outreach and public education activities, the HRF reaches more than five million people a month with accurate, positive information about the health benefits and safety of herbs. The HRF's membership now includes thousands of individuals and companies in forty different nations around the world. HRF president Rob McCaleb comments, "I'm proud of the Herb Research Foundation, of the dedication of its staff and members who have built it into a leading voice for the responsible use of herbs in modern health care."

Background

From its modest beginnings as a two-person office, the HRF now employs a core staff of seventeen plus a number of work-study employees and volunteers in a 2,900-square-foot office space overlooking the Pearl Street Mall in Boulder, Colorado. Its library currently contains more than 200,000 scientific articles on approximately 1,500 plants, and it maintains a proprietary database that is continually updated with the latest evidence supporting the use, benefits, and safety of more than 200 commonly used herbs.

McCaleb envisioned the HRF during his fourteen years as research director at Celestial Seasonings, Inc. For the first few years, the HRF operated out of the Celestial offices before moving to an independent location. The relationship with Blumenthal was based in part on an agreement to copublish the peer-reviewed journal, *HerbalGram,* for which McCaleb continues to serve as technical editor. The first meeting of the HRF Professional Advisory Board, made up of eighteen of the world's leading medicinal plant experts, was held in October,

1982. In 1989, the HRF took up residence at its present location at 1007 Pearl Street, Boulder, Colorado.

As the HRF has grown, so has the scope of its information services and public outreach activities. Drawing from its extensive library and multiple online databases, the HRF provides custom botanical literature research services, Herb Information Packets on herbs and health conditions, and the popular Natural Healthcare Hotline service. The HRF Web site at www.herbs.org features an informative mix of hot news, research reviews, and an interactive "speak out" page where visitors can post questions to be answered by HRF information specialists. The HRF editorial department manages the writing, editing, and peer-review process for the Research Reviews department of the quarterly journal, *HerbalGram,* published jointly with the American Botanical Council, and containing research reviews, legal and regulatory updates, and in-depth articles on medicinal plants. *Herbs for Health,* founded in 1996, is published six times a year; it offers basic, practical information on using medicinal herbs.

Objectives

The HRF's primary mission is to foster improved world health and well-being through the informed use of herbs:

1. To encourage and support research into the benefits of herbs, and to establish a solid scientific foundation for their use as foods and medicines
2. To serve as a reliable source of information on the safe and appropriate use of medicinal plants and herbal products for the public, professionals, legislators, and the news media
3. To publish and disseminate accurate research information to researchers and the public
4. To promote the rational, informed use of herbs in preventive medicine to improve and maintain health and lower health care costs
5. To promote and participate in programs and activities that return serious medicinal plant study to prominence at American universities and medical research facilities
6. To support developing countries in their pursuit of rational, cost-effective health care with natural remedies, and to help preserve traditional health care systems through identifica-

tion, preservation, and cultivation of important plant remedies threatened by overcollection

7. To raise awareness about and protect threatened medicinal plant species in the United States and abroad, and to help save undiscovered medicinal plants from extinction
8. To form a liaison between the American herbal movement and the worldwide scientific community
9. To increase the sustainable cultivation of herbs as hardy cash crops for farmers in both developing and developed countries
10. To bring natural product medicines back into legitimate use in modern health care

The HRF receives no public funding but depends on the support of its members.

Through its media outreach and education programs, the HRF reaches millions of people each month with accurate information on the safe and appropriate use of herbs. It is involved in sustainable herb cultivation projects throughout the world, which help improve the quality of life for people in developing nations while protecting wild medicinal plants endangered by overcollection. The HRF services include custom botanical literature research, online document delivery, publications, third-party educational content, and Herb Information Packets.

Another important area for the HRF is sustainable herb development, which fosters a number of vital goals: bolstering rural economies and improving quality of life for disadvantaged farming families, protecting local medicinal plant populations by enhancing access to cultivated herbs, and providing a source of high quality, organic botanicals for the international market. For example, the HRF's Africa project touches the lives of hundreds of farming families by boosting local economies with earth-friendly agriculture and providing communities with training, jobs, and better access to low-cost botanical medicine.

Under a three-year contract with the U.S. Agency for International Development's Africa Bureau and the U.S. Department of Agriculture (USDA), the HRF is now working with the Agriculture Research Council of South Africa to assess opportunities for botanical agribusiness and assist in improving local agricultural research and business development networks in South Africa. If successful, similar projects will be implemented in other sub-Saharan countries. In 2000, the HRF cosponsored the first of a planned series of international development symposiums designed to provide participants with the tools they need to find a profitable niche in the worldwide botanicals market.

BIBLIOGRAPHY

"Celebrating 50 Issues of *HerbalGram*." *HerbalGram 50* (2000): 10-11.

Griggs, Barbara. *Green Pharmacy: The History and Evolution of Western Herbal Medicine*. Rochester, VT: Healing Arts, 1997.

Herb Research Foundation Home Page: www.herbs.org. Material from the Herb Research Foundation Web site reprinted with permission of Robert S. McCaleb, President.

HERB SOCIETY OF AMERICA

Established in 1933, the Herb Society of America is dedicated to furthering the knowledge and use of herbs, and to contributing the results of the experience and research of its members to the records of horticulture, science, literature, history, the arts, and economics. It is also concerned with protecting the global environment for the health and well-being of humankind and all growing things. The Society has adopted herbalist John Parkinson's phrase, "For Use and For Delight" as its motto.

Membership is open to any individual or group interested in herbs. According to the Herb Society of America home page, more than 3,000 members enjoy the following benefits:

- Access to the library (with more than 2,000 volumes alone) for research and information
- Subscription to the annual journal, *The Herbarist,* devoted to the latest findings, writings, and information on herbs; the quarterly newsletter; and the annual report
- Discounts on books purchased through the Garden Gallery Gift Shop as well as subscriptions to *The Herb Companion* and *Herbs for Health*

- Special rates on Society-sponsored regional and national symposia
- Access to the national conference, which features educational tours, workshops, and lectures
- Networking opportunities with authors, lecturers, researchers, businesspeople, and expert gardeners
- Support of the Seed Exchange Program and the National Herb Garden

Research and Outgrowth

The Herb Society of America provides funds to support the National Herb Garden located on 2.5 acres at The National Arboretum, Washington, DC. The purposes of the National Herb Garden include displaying herbs for the education of the public and studying herbs for their potential use. Its library, located at the National Headquarters of The Herb Society of America, is open to the public five days a week for research. A staff librarian is available to assist people in search of information. Over 500 inquiries are made to the Library each year in person or by telephone.

In addition to *The Herbarist,* HSA publications include *The Herb Society of America 1933-1993: A History,* by Dorothy Spencer, *Traveler's Guide to Herb Gardens,* and *Judging Herbs: A Handbook for Horticultural Judges*. The Herb Society of America also endorses *The Encyclopedia of Herbs & Their Uses,* by Deni Bown. All publications are available from the National Headquarters.

Seed Exchange

This is one of the many educational activities of The Herb Society of America involving botany and horticulture. Members volunteer their time to collect, package, and catalog seeds. The selection varies year to year but includes common and lesser known culinary, aromatic, economic, and medicinal herbs plus many varieties commonly grown as ornamentals. Seeds are available at a nominal price from March to early December.

At The Herb Society of America's headquarters, there is a room for children to sit down to a story program on herbs, educational displays on herbs that greet guests as they enter the front gallery, a greenhouse and classroom for children to experience hands-on the joy of herbs, a horticulturist who gives advice to a gardener on how to grow sage, and distance learning programs in partnership with area colleges that inspire and connect people to herbs worldwide. Several members of a visiting Canadian unit explore new program ideas to take home to create in their community. In a serene rural setting, surrounded by The Holden Arboretum and Metroparks, visitors from around the world come to learn, to enjoy, and to discover herbs.

Plant Collections

In 1996, The Herb Society of America established a series of Plant Collections held by units or member groups within the organization.

Purpose

HSA recognizes that living herb plants are the very reason for our existence and, therefore, encourages the cultivation and preservation of the living germplasm of herbs. These national collections of herbs serve a role in:

- Education, through exposing our members and the general public to a wide variety of herbs
- Research, by making these collections available for study
- Exchange, by sharing a limited number of cuttings, seeds, or plants of otherwise unavailable herbs through The Herb Society of America Seed Exchange and commercial sources
- Conservation, by encouraging the cultivation, reintroduction, and preservation of uncommon herbs for posterity and by serving as a repository of authentically labeled taxa

About Collections

Collections may be generic (e.g., Lavandula), a division within a genus (e.g., hybrids bred in the United States), cultivars within a species (e.g., all the cultivars of *Rosmarinus officinalis*), native species only (perhaps focusing on rare ones), geographical (e.g., Asian herbs), and uses (e.g., dye and fabric herbs). Parallel, complementary, or duplicate collections are desirable to cater to our range of climates and to encourage sharing between and among collection holders. In addition, growing three plants of each taxon in any one collection ensures that enough plant material is available. The association of a collection with a commercial nursery may be desirable.

Provision of Propagating Material

Plant material from the collection may be sold, exchanged, or donated at the discretion of the collection holder. The collection holder should be sensitive to the distribution of potentially invasive, exotic species.

Research Grants

The Herb Society of America has awarded over $75,000 to the advancement of promising research. Previous grants have researched the archeology, history, taxonomy, physiology, genetics, and tissue culture of herbs. Its purpose is to further the knowledge and use of herbs and contribute the results of the study and research to the records of horticulture, science, literature, history, art, or economics. Past award winners include:

1970-76	Genus *Mentha*
1977	Filming of vernacular medical manuscript in Folger Library
1977-78	Plant lists for the National Herb Garden, National Arboretum
1979	Completion of plant lists for the National Herb Garden, National Arboretum
1980	Transcription of vernacular medical manuscript in Folger Library, *Louisiana Kitchen Gardens, 1750-1850*
1981	Study of the taxa of Thymus
1982	Susceptibility of herbs to air pollution; transportation of collection of Mentha herbarium from Michigan to Delaware
1983	Final typing of manuscript: *This Booke of Sovereigne Medicines Handbook on Gardens: Archeological Evidence*

1984	Plants in the fourteenth and fifteenth century: Paintings of St. Sebastian
1985	*A Weaver's Garden*
1986	*Herbs in the Landscape: A Design and Planting Guide*
1987	Cell culture studies of pokeweed antiviral compound; on-site research of medieval plants and gardens in Hungary and Poland
1988	Tussie-mussies, nosegays, and sweet poseys; a taxonomic study of the genus *Ocimum*
1989	Tropical herbs of Hawaii; printing of three color plates for the Fuchs Herbal; genetics of sex allocation in coriander
1990	Imagistic associations of garlic in Greek and Latin literature; landscape uses and hardiness of herbs in Tennessee
1992	The inactivation of lipoxygenase and peroxidase activities by extracts of herbs; *Herbalist to the King: The Travels and Writings of Philibert Commerson*
1993	*Rascally Verses and Rougish Tobacco: The Tobacco Polemic in Early Modern England;* winter survival and effects of pruning on Lavandula species and cultivars; identifying genetic markers and their role in selecting chemotypes in perennial Lamiaceae
1994	Herbs as inducers of protective proteins
1995	Development of a DNA Probe for the detection of Fusarium in basil
1996	Molecular biological analyses of herb action; a high intensity indoor herb garden for lunar and planetary colonies; dramatic poisonings in Elizabeth England
1997	Traditional dye plants of the Huichol Indians of Mexico
1998	No grant awarded
1999	Localization and Conservation of Genetic Variation in Ginseng *(Panax quinquefolius):* An Imperiled American Herb
2000	Plant representations in byzantium—A pictorial pictionary; preserving knowledge of traditional plants in the Houma Community of Louisiana
2001	Phylogenetic relationships and taxonomic status of narrowly endemic Monarda taxa

The organization is based in Kirtland, Ohio, near Cleveland. The headquarters building was built in 1841 utilizing local stone by a

Lake County farmer, Henry Earl. Believed to be the oldest stone structure in the county, it is on the National Registry of Historical Places. Open for visits, the address is 9019 Kirtland Chardon Road, Kirtland, Ohio 44094; phone: (440) 256-0514; fax: (440) 256-0541.

BIBLIOGRAPHY

Bown, Deni. *The Herb Society of America Encyclopedia of Herbs and Their Uses.* London: Dorling Kindersley, 1995.

The Herb Society of America Home Page: www.herbsociety.org. A wide range of links are available, including "History & Purpose," "Membership," "Publications," "Projects & Programs," and "Staff and Contact Information."

HERBAL ASPIRIN

The Evolution of Aspirin

Centuries before the discovery of aspirin, extracts of poplars and willows were widely employed as analgesics. In time, the active ingredient was ascertained and named salicin after the genus of willows, *Salix.* However, salacin possessed a bitterness that could not be sweetened, was insoluble in water, and frequently led to stomach and intestinal irritation and even bleeding. Aware that salicin was converted to the sweetish-tasting and less irritating salicylic acid in the stomach, scientists isolated the latter compound from the flower buds of meadowsweet *(Filipendula ulmaria)* in 1939. Later, they found ways of producing salicylic acid out of the salicin extracted from other sources, most notably willow and poplar bark.

Many physicians in the latter half of the nineteenth century prescribed salicylic acid even though it occasionally irritated the stomach. A breakthrough occurred in 1899 when the Bayer pharmaceutical company (based in Germany) synthesized acetylsalicylic acid and began manufacturing it under the brand name of Aspirin (a reference to meadowsweet's scientific name, *Spiraea ulmaria*). Considerably less harsh on the digestive tract, aspirin proved so popular that it became a household term overnight. Due to Bayer's failure to adequately trademark the name, competitors soon capitalized on aspirin's popularity, forcing prices down and spreading its use even further.

Herbal Offshoots of Aspirin

Salicylic acid remains the basis of the aspirin group of drugs. Along with related analgesic compounds such as methyl salicylate and the aforementioned salicin, it has long been used as an alternative to pharmaceuticals for relieving pain, lowering fevers, reducing inflammation, and even overcoming mild insomnia. Wild analgesics can be derived from a wide range of plants, including

- The cambium (inner bark) of many varieties of willows and poplars
- All birches, particularly the black or sweet birch
- Queen of the Prairie *(Filipendula rubra)*
- Wintergreen (also known as teaberry)
- Yarrow
- Black cohosh
- Pennyroyal

The analgestic features of these plants are generally used medicinally as either a tea or essential oil. Essential oils of wintergreen and sweet birch are available from specialty retailers. Highly concentrated (containing up to 99 percent methyl salicylate), they can, if used inappropriately, cause significant internal or external irritation. The tea extracts—available both commercially and medicinally—are named according to the mode of preparation: an infusion or tisane means boiling water has been poured over the herb and allowed to cool and steep; a decoction indicates the herb has been placed into water and simmered for some time. As with external application of essential oils, medical experts recommend caution when ingesting these teas because the plants from which they are derived contain many compounds other than the salicin family. The strength of the tea should be based upon desired effect combined with attention to the messages sent by the body following ingestion.

BIBLIOGRAPHY

Dominick, Annie Westbrook. "Herbal Aspirin; Headache and Fever Remedies *Do* Grow on Trees." *The Herb Companion* (August/September 1992): 36-39.
Griggs, Barbara. *Green Pharmacy: The History and Evolution of Western Herbal Medicine.* Rochester, VT: Healing Arts, 1997.

HERBAL COSMETICS

Cosmetics based on herbs and other botanicals are at least as old as civilization itself. Bath oils and rubs, moisturizing and cleansing lotions for the skin, shampoos and conditioners for the hair—these and other products utilize the oils and other by-products of herbs both as central ingredients and subtle additives.

Egypt, the cradle of one of the earliest ancient cultures, pioneered natural perfumes and skin preparations. Sarah Garland states that "Egyptian ladies washed in scented waters, outlined their eyes with kohl, used an eye shadow of powdered lazuli and a rouge of henna leaves, and all Egyptians kept their skins supple with oil rubs; the rich using precious perfumed oil, the poor a crude palm oil" (p. 244).

The Roman poet Ovid writes that the women of his day used a mixture of barley, bean flour, eggs, hartshorn, pulped narcissi bulbs, honey, and aromatic gums to create face packs for whitening the skin; he further notes, "every woman who spreads this on her face will render it smoother and more brilliant than her mirror." The Roman baths featured water scented with lavender and other aromatic herbs; bathers capped their sessions with massages incorporating olive oil and scented oils such as saffron oil.

Asian cultures placed a comparable premium on herbal cosmetics. Most Eastern women have traditionally scented their hair with jasmine and frangipani flowers, while both sexes widely use henna (for red tints), saffron (yellow), and other plants for dyeing hair. Garland notes that "Indian women . . . used (and still use today) . . . kohl around their eyes and henna on hands and feet, and their baths were scented with heavier perfumes, jasmine, patchouli, and sandalwood" (p. 244).

A watershed development occurred in the tenth century when an Arabian physician, Avicenna, discovered how to distill the essential oil from plants. Herbs had become so important to the lives of Europeans by the early sixteenth century that most large houses possessed a still-room; a portion of every garden was designated for "herbes to stylle."

The advent of motion pictures and mass-market advertising in the twentieth century created new perceptions of feminine beauty and greater demands for cosmetics. Many consumers found advantages to making cosmetics at home, most notably the opportunity to employ pure and natural ingredients, customized colors and scents, and preparations suitable for varying types of skin and hair. The box provides an inventory of ingredients frequently used to make simple cosmetics in the home.

Basic Ingredients for Homemade Cosmetics

Herbs for Skin Care

- Borage
- Burdock
- Camomile
- Chervil
- Coltsfoot
- Comfrey
- Dandelion
- Elder
- Eucalyptus
- Fennel
- Goosegrass
- Horsetall
- Houseleek
- Lady's Mantle
- Lavender
- Lemon balm
- Lime
- Lovage
- Marigold
- Marshmallow
- Mint
- Parsley
- Plantain
- Rosemary
- Sage
- Stinging nettle
- Thyme
- Witch hazel
- Yarrow

Supplementary Ingredients

- Alcohol
- Alum
- Arrowroot
- Beeswax
- Benzoin or gum benjamin
- Borax
- Cocoa butter
- Fuller's earth
- Glycerine
- Honey
- Kaolin
- Lanolin
- Lemon juice
- Oil
- Vinegar
- Water

Source: Garland, 1993, pp. 246-247.

The chief categories of cosmetics include skin care preparations; perfumes, including scents derived by distillation, extraction, enfleurage, and maceration; bathing products; soaps; and shampoos and conditioners. All of these categories are well represented by herbal recipes. For example, popular herb scents include hungary (fresh, crushed rosemary; fresh, crushed mint; crushed rose petals; grated lemon peel; orange-flower water; vodka) and cologne oil (bergamot oil, neroli oil, sweet orange oil, lemon oil, lavender oil, rosemary oil, petitgrain oil, diethyl phthalate).

BIBLIOGRAPHY

Garland, Sarah. "Cosmetic Herbs." In: *The Complete Book of Herbs & Spices.* Pleasantville, NY: Reader's Digest Association, 1993, pp. 244-261.

McHoy, Peter, and Westland, Pamela. *The Herb Bible.* New York: Barnes & Noble Books, 1994.

HERBAL EDUCATION

The herbal renaissance of the late twentieth century saw not just a heightened interest in and use of herbs, but an increased demand for opportunities to learn more about herbs. Writer Kate Carter Frederick states, "Traditionally, herbal studies have taken two routes: live and learn with the plants or glean anecdotal wisdom from the experience of others" (p. 58). At present, a wide range of herbal education opportunities are available depending on one's age, attitude, financial situation, lifestyle, and locale. Directories of education providers are available from the American Herbalists Guild (Soquel, California) and the Intra-American Specialties (West Lafayette, Indiana), among others. Such groups offer the only educational guidelines relating to herbal education. However, many formal programs do provide certification once prescribed course work is completed.

It is helpful for anyone interested in herbal education to begin by setting goals. When considering available options, Frederick recommends asking the following questions:

- What is the curriculum or schedule of events?
- What area of herbal studies is addressed? What approach is employed?
- For whom is the program designed?
- What is the method of instruction?
- What are the instructor's credentials?
- Are there learning opportunities beyond those designated?
- What is the overall cost of the program?
- What resources are available (e.g., textbooks, crafting ingredients)?
- Is the location convenient?
- Is the schedule manageable?
- Is certification offered?

The following are among the most popular learning options:

Herb gardens and walking tours. According to New York herbalist and teacher Susun S. Weed, "Hands-on is the best way to learn." Settings include yards tended by herb hobbyists, herb farms, and public gardens and greenhouses.

Herb courses, workshops, retreats. A Vermont-based educator, Rosemary Gladstar, has stated that "the best way to learn about herbs is by taking a class." Workshops place more emphasis on "doing" than structured classes; they also offer the greatest diversity of topics and settings. Retreats generally include herb walks, classes, projects, demonstrations, conversations, and meals.

Herb schools. Two major types of ongoing educational programs exist: the workplace (sometimes the home) of an herbalist and traditional schools. While each institution may possess its own style and approach, typical course topics include herbal preparations and therapies, history, human anatomy and physiology, botany, ecology, and field activities.

Seminars, symposiums, conferences. These options all provide a platform for prominent professionals to share their fields of expertise with an interested audience. Seminars are usually one- or two-day sessions geared to advanced students. A symposium is frequently a weekend event embracing speeches, panels, and workshops covering varied topics centered around a broad theme. Conferences are larger in scale, including speakers, forums, workshops, a trade show, organizational meetings, outside tours, and other special events.

Apprenticeship. Formal apprenticeships generally run two to twelve months in length. Activities might include assisting in research and workshops, acting as a tour guide, tending gardens, harvesting herbs, packaging products, and managing business operations.

Correspondence courses. Commonly employed as a complement to other learning modes, correspondence courses are ideal for self-motivated students unable to attend school full time. Up to the mid-1990s, the typical format consisted of students receiving lessons incrementally by mail, and then returning assignments to the instructor for evaluation, feedback, and individual guidance. E-mail and electronic classrooms complete with chat rooms, Web sites, and other distance learning tools have taken over since then.

Herb study groups, clubs, professional organizations. In return for basic dues, many of these organizations offer camaraderie, travel, regular meetings, special educational events, newsletters and discounted publications, and resource databases (including membership lists).

Print and electronic media. The herbal renaissance of the late twentieth century spurred the publication of a plethora of magazines and books. The latter include field guides, herb encyclopedias or compendiums, reprints of classic herbals, cookbooks, and specialty books on such topics as cosmetics, crafts, and the business applications of herbs. The herb video market is expanding rapidly and the Internet has made far greater quantities of information available to the public at large.

By the mid-1990s, the Internet offered a variety of additional educational opportunities. For instance, NaturalHealers.com provides information on schools, programs, certifications, and licensing requirements for those interested in careers as massage therapists, naturopathic doctors, chiropractors, acupuncturists, and other healing professions.

BIBLIOGRAPHY

Frederick, Kate Carter. "Herbal Education." *The Herb Companion* (August/September 1994): 58-62.

NaturalHealers Home Page: www.naturalhealers.com. NaturalVillage.com, 2000.

HERBAL INCENSE

The burning of tree barks and saps, gums, resins, roots, flowers, fragrant leaves, and needles to provide a fragrant smoke may well go back to man's early use of fire. An abundance of references to incense in the Old Testament indicates that the Hebrews have used it since the early days of civilization while the ancient Egyptians burned offerings to the sun-god, Ra. The Greeks of antiquity employed incense to render sacrifice and prayer more acceptable to their deities.

Incense had been incorporated into most Christian rituals (e.g., high mass, processions, funerals) by the fourteenth century. While incense was unknown to early Buddhism, with its opposition to external dogma, Tibetan, Japanese, and Chinese Buddhists now use it widely in both public and private settings. Modern Hindus regularly burn camphor and incense to Krishna's image.

Modern pagan and neopagan practitioners have developed their own ritual frameworks employing incense. Native American religious rituals have long featured the burning of cedar, sage, sweet grass, tobacco, uva ursi, and yerba santa for purifying oneself and the surrounding environment, for praying to the Great Spirit, and for connecting with one's spirit helpers.

Sandy Maine notes that incense can also be used to "evoke a mood or create an atmosphere for shopping, entertainment, romance, or home relaxation. It's a mental stimulant that can bathe ordinary events and activities in a special glow" (p. 36).

Incense is used in a variety of forms, ranging from loose ingredients (typically thrown on glowing coals) to dough-based formulas which can be shaped into cones, cylinders, sticks, or coils. With the exception of the loose form, incense generally includes the following ingredients:

- An aromatic substance or mixture; preferred materials are herbs, spices, botanical powders, woods (e.g., sandalwood, juniper), and bark (cinnamon).
- The base, a substance that aids in the burning of the aromatic, frequently enhancing or tempering the scent. Most popular are

powders derived from woody plants such as cassia, charcoal, evergreen needles, sandalwood, vetiver, and willow.

- The bonding agent, a resin or gum that holds the aromatic and base together (ideally burning well without giving off toxic smoke); agar, gum arabic, karaya, and tragacanth are most widely used.
- A liquid for changing the bonding agent into a glue. While water is the cheapest and easiest to use, herb waters, brandy, wine, olive oil, and tinctures also enjoy favor (despite the lack of any significant difference in the odor or burnability of the incense).
- A coloring agent; although food coloring may provide the most convenient option, plants can supply natural colors (e.g., willow for brown, safflower for yellow).

BIBLIOGRAPHY

Fettner, Ann Tucker. *Potpourri, Incense, and Other Fragrant Concoctions*. New York: Workman Publishing, 1977.

Maine, Sandy. "Herbal Incense." *The Herb Companion* (December/January 1992/1993): 36-40.

Smith, Steven. *Wylundt's Book of Incense*. York Beach, ME: Samuel Weiser, 1982.

HERBAL MARKETS AND CONSUMERS

The commercial herbal industry began from two diverse segments of the U.S. population: the back-to-the-land, home-remedy folks who came out of the late 1960s and those who embraced a natural lifestyle as part of their heritage and developed an industry based on natural remedies. Today, consumers are interested in returning to a more natural style of life and they are dissatisfied with modern medicine when they perceive it as ineffective, expensive, and carrying unwanted side effects. There is an increasing awareness of health and fitness among aging baby boomers seeking to prolong their good health and active lifestyle into old age.

Conversely, there are also reasons for why the industry has not grown more rapidly, such as the lack of cultural understanding necessary to incorporate herbal products into daily life. The United States is behind other countries in its use, understanding, and regulation of botanical products and has, in fact, been labeled a "third world country" when it comes to the use of botanicals. Also, growth has been hampered by a regulatory climate that disallowed any health information to accompany the product.

In 1993, the *New England Journal of Medicine* published a survey on the use of unconventional medicine in the United States (Brevoort, 1996). According to its findings, 34 percent of all Americans reported using some type of unconventional medicine in the previous twelve months. Of this, 3 percent said they had used herbal medicine and 10 percent saw a provider of herbal medicine. Total expenditures for unconventional therapy amounted to approximately $13.7 billion, of which three-quarters was paid out-of-pocket—herbal remedies are not reimbursable as part of insurance coverage today.

In general, the U.S. herbal market is difficult to assess for a number of reasons. First, the industry is in an expansion mode and it is difficult to get a handle on the change. The market is made up of a few large and many small closely held companies who consider their financial information highly proprietary and do not willingly share it. No trade group has compiled reliable statistics on their members' sales. The industry has been too small to afford to do its own market research.

Medicinal Herb Industry

This is a very specialized group that has grown almost 300 percent since 1992. Many large companies that historically have sold mainly vitamin supplements in mass market stores—most notably, Leiner Health Products, Pharmavite, and Nature's Bounty—are committed to a line of herbal products. These products now constitute a majority of vitamin sales (52.5 percent). The category of medicinal herbs is not clearly broken down by the Department of Commerce. Herbal powders may not be labeled to indicate they will go into the medicinal market, and the customs value is underestimated; this is particularly true for wild Chinese ginseng *(Panax ginseng),* the most expensive botanical in the world. Unofficial industry estimates state that 75 percent of the botanicals used in medicinal products are imported and 25 percent are domestically grown.

The main herbs that are grown domestically are echinacea, goldenseal, hops, peppermint, slippery elm, valerian, and aloe. Of these, echinacea, goldenseal, and slippery elm are wildcrafted as well as cultivated. Ginkgo, echinacea, and peppermint are commercially grown in the United States and saw palmetto is wildcrafted and exported to Europe for manufacture into advanced phytomedicines. American ginseng is exported to the Orient. In 1994, almost 2.4 million pounds of American ginseng were exported with a declared value of $76 million. The United States has become a commodity supplier for the industrial world market for processed herbal medicines.

Chinese Herbs

Chinese herbs are a special subcategory of the American herb market. In 1980, Chinese herbs were rarely seen outside of the Oriental ethnic community. More than twenty years later, they are imported as packaged products from the Orient ("patent medicines") and sold in natural food stores by health practitioners trained in their usage. Most Chinese herbs are used in formulas rather than as single herbs. Aside from the retail sales of these formulas, many are dispensed by acupuncturists, and by naturopathic physicians, who are licensed to dispense herbs in at least twelve states.

Herbs of Commerce

Because of the growing need for guidelines in the sale of botanicals, the American Herbal Products Association compiled a comprehensive review of the traditional literature on 600 of the most widely sold herbs in America from their journal, *Herbs of Commerce,* with the botanical and common names of 600 herbs on the market; the list has now been expanded to include over 1,800 herbs.

Market Trends

With the advent of a new millennium, several trends have developed in the U.S. herbal market. There is an increased emphasis on high-quality products and on scientific validation for products in the marketplace; more liberal regulations of botanicals; more emphasis on the phyto-pharmaceutical model (based on standardized active ingredients within an herbal framework); increased interest in nutraceuticals (foods containing ingredients that have been scientifically proven to improve health); more mainstream companies entering the marketplace; and more funded research on the long-term health benefits of botanicals as well as their curative properties. With the U.S. herbal dietary market growing at an annual rate of 25 percent and reaching four billion dollars based on late 1999 estimates (compared to six billion dollars in Europe), some experts feel the saturation point is imminent.

BIBLIOGRAPHY

Blumenthal, Mark. "Herb Sales Down 3 percent in Mass Market Retail Stores, Sales in Natural Food Stores Still Growing, but at Lower Rate." *HerbalGram 49* (2000): 68.

Brevoort, Peggy. "The U.S. Botanical Market: An Overview." *HerbalGram 36* (Spring 1996): 49-57.

Galinsky, Bob. "Imported Herbs Gaining in Popularity." *Tropical Produce Marketing News* (December/January 1995/1996): 1p. www.fintrac.com/gain/marketstats/tpm.

"US Herbal Market Nearing Saturation." *Herb World News Online.* Boulder, CO: Herb Research Foundation, 2000. 1p. www.herbs.org/current.

HERBAL PILLOWS

Pillows stuffed with herbs have been used for centuries as an aid to bringing on sleep. Rose petals were a favorite of the ancient Romans. England's King George III and Abraham Lincoln allegedly used hops-based filling. Victorian era women employed small lacy pillows filled with lavender and rose petals. Even the Bible includes references to this practice; e.g., the presence of Our Lady's bedstraw *(Galium verum),* a sleep-inducing herb, in the cradle of the baby Jesus.

The role of herbs in inviting slumber is founded in scientific fact. Researchers have explored the ability of lavender oil to make laboratory mice drowsy. Aromatherapists have long argued that fragrance acts as a stimulus for significant physiological and psychological reactions. As Kathleen Halloran notes,

> Because olfactory neurons are connected directly to the brain, simply inhaling a scent can stimulate the release of hormones that generate a range of feelings and responses; they can calm fear or anger, relieve stress or pain, and yes, bring on sleep. There is thus foundation for the belief that a pleasant fragrance in one's pillow can alleviate mental and emotional stress, a frequent cause of sleeplessness. (p. 47)

While herb pillows can come in any size or shape, traditionally they have been small and relatively flat so they fit easily within a standard pillowcase or under a bed pillow. Furthermore, smaller pillows are more adaptable as attractive accent pieces and for storage in closets and drawers where they can scent other linens. Generally, the pillow (which can be augmented with ornamentation—e.g., lace, ruffles, ribbons, etc.—according to personal taste) should serve as an envelope for holding a separate herb bag made of muslin, tulle, or similar thin material inserted in the back. The outer pillow should be washed as needed, and the herb bag replaced (or replenished) after the fragrance has faded.

Herbal ingredients can likewise vary by individual preference. The key consideration is the use of pleasant associations as an aid to falling asleep. Fragrances associated with childhood (e.g., sweet woodruff and sweet clover evoking the smell of grass on a summer day) seem to work well in this capacity. Other herbs are prized for their calming (rosemary leaves, peppermint leaves, lemon-scented plants, etc.) or sedative (lavender flowers) effects. Herbs can also be mixed in the manner of potpourris utilizing a fixative such as benzoin or orrisroot to achieve a longer-lasting fragrance. In addition to the aforementioned herbs, the following (they usually work best in dried form) have proven successful as ingredients in herb pillows: sweet marjoram, thyme, lemon verbena, lemon balm, chamomile flowers, eucalyptus, pine needles, crushed aniseed, bergamot oil, hop bracts, ground allspice, orange peel, neroli (orange blossom) oil, spearmint leaves, orange mint leaves, ground cloves, and ground tonka bean.

BIBLIOGRAPHY

Garland, Sarah. *The Complete Book of Herbs & Spices*. Pleasantville, NY: Reader's Digest Association, 1993.

Halloran, Kathleen. "Sleep Pillows." *The Herb Companion* (December/January 1993/1994): 47-49.

McHoy, Peter, and Westland, Pamela. *The Herb Bible*. New York: Barnes & Noble Books, 1994.

HERBAL SOAP

The use of soap as a cleaning agent probably reaches back to the beginning of civilization. Written references to soap were found on Sumerian clay tablets in Mesopotamia (now southern Iraq) dated around 2500 B.C. It was used widely throughout the Roman Empire. The excavation of Pompeii, covered by the volcanic eruption of Mount Vesuvius in A.D. 79, indicated the presence of a full-fledged soapmaking establishment. The word itself came from Roman legend. Rain washed a mixture of melted animal fats and wood ashes from Mount Sapo (where animals were sacrificed) to the Tiber River below; the resulting soapy residue was found to be of value in cleaning both clothing and the body.

The fall of the Roman Empire in the fifth century led to a decline in bathing and the use of soap. However, soapmakers' guilds began appearing in Europe in the seventh century. Italy, Spain, and France were leading producers due to the availability of olive oil and barilla, saltwort, and other plants whose ashes proved useful in manufacturing lye. Throughout the Middle Ages, soap was heavily taxed as a luxury and, therefore, readily available only to the upper classes.

The high cost of soap encouraged women in Colonial America to make it at home just as they did most other domestic products. While commercial production of soap had begun in 1608, when several English soapmakers established themselves in the New World, the American industry didn't truly flourish until the mid-1800s with the implementation of technological advances such as a process for extracting inexpensive soda ash from common salt.

The general public's fascination with domestic crafts spurred a resurgence in the popularity of home soapmaking in the mid-1980s. Sandy Maine, founder and owner of the Sunfeather Herbal Soap Company, estimated that at least 6,000 home hobbyists and 150 small commercial soapmakers existed a decade later. Many soapmakers, both at home and in the marketplace, were actively incorporating herbs and herbal scents into their products. These ingredients can add color, fragrance, texture, and gentle soothing action to soap. Preferred herbs for soap recipes include calendula, comfrey, aloe vera, rosemary, lavender, peppermint, spearmint, lemon verbena, lemongrass, chamomile, cedarwood, cinnamon, dill weed, coconut oil, and turmeric.

BIBLIOGRAPHY

Garland, Sarah. "Herbs for Washing and Polishing." In: *The Complete Book of Herbs & Spices*. Pleasantville, NY: Reader's Digest Association, 1993, pp. 224-225.

Maine, Sandy. *Soap Book*. Loveland, CO: Interweave Press, 1995.

Maine, Sandy. "Soaps: Scented Essentials." *The Herb Companion* (October/November 1995): 50-55. (Includes a directory of companies offering soapmaking supplies or kits, a sampling of recipes for herbal soaps, and a sidebar entitled "How to Make Soap.")

HERBS AND PETS

The practice of providing herbs for pets has gained increasingly widespread popularity in recent decades as a means of amusement in addition to making their lives more comfortable. Many pet owners have found commercially available pet foods such as dog biscuits to be bland in taste and lacking in the nutrients available in domesticated herbs. Likewise, stuffed toys containing catnip lack the freshness and pungent fragrance that can be derived from homegrown herbs.

Christine Wittmann, in the Winter 1990 issue of *The Herb Quarterly,* has provided practical instructions in the preparation of many herb-based treats for family pets. They include:

- Kitty mats composed of a cotton cover, cotton batting (or fiber filler), and dried, finely crushed catnip leaves
- Catnip-filled mice toys for cats utilizing soft wool, felt, or pillow ticking
- A walnut-sized portion of valerian root tied in a sturdy piece of cloth (e.g., calico) to be used as a feline plaything
- Flea collars (either cotton lamp wick saturated with herbal oils such as pennyroyal, citronella, and rosemary, or a cloth sack containing a mixture of dried herbs such as cedar shavings, roman wormwood, southernwood, pennyroyal, mugwort, and lemon verbena) or other repellents (stuffed pillows, treated bed covers, etc.)
- Flea powder containing pennyroyal, roman wormwood, eucalyptus, lemon verbena, sassafras root bark, and diatomaceous earth
- Dog biscuits incorporating parsley flakes, fresh chopped chives, dried chopped lovage leaves, and finely chopped garlic clove
- Cat crunchies utilizing dried or powdered alfalfa herb, dried parsley, dried stinging nettle herb, dried horsetail herb, and freshly grated garlic
- Portable herb gardens (i.e., cultivated seeds of wheat grass, oat grass, catnip, lemon catnip, etc.) housed in shallow containers with potting soil for indoor cats.

BIBLIOGRAPHY

Lawson, Tony. *The Cat-Lovers' Cookbook.* Illustrated by Pate Lawson. Pownal, VT: Storey Communications, 1994.

Wittmann, Christine. "Herbal Treats for Pets." *The Herb Quarterly* (Winter 1990): 12-16.

HERBS IN WEDDING CELEBRATIONS

Herbal decorations have been a part of the marriage ceremony for centuries. A wedding celebration honors the union of two lovers and observes the customs passed down through generations. Champagne toasts, the white dress, and the nuptial vows are all symbols carried over from the past. One of the most treasured wedding customs is the use of herbs to enhance the occasion. With their pleasing scent, beauty, and folkloric associations, herbs and flowers bring a wealth of tradition to a wedding table. From the communal feasts of the Middle Ages to the high-society fetes of the nineteenth and early twentieth centuries, wedding herbs have been a fixture in one of mankind's central traditions.

Medieval Wedding Herbs

Brides in medieval Europe were well versed in herbal lore. Their herb gardens functioned as both pantry and pharmacy for immense households of families and servants. At weddings, herbs were valued most for their aromatic properties and simple decorative uses. Most popular was rosemary, the ancient plant symbolizing memory and friendship, which graced many a medieval ceremony with its aroma and flavor. At banquets, court rituals, and religious rites, festive garlands of fresh rosemary brought warmth and life to cold stone walls of castle and church. Rosemary's sharp, clean scent was heightened by incense braziers, a tradition inherited from the ancient world. Wedding customs surrounding rosemary stressed its association with marital fidelity. At some ceremonies, the bridal couple dipped a rosemary sprig into each other's wine before toasting their love. Other wedding parties presented gilded branches of rosemary tied with colorful ribbons as favors to wedding guests, symbolizing the bridal couple's new life.

Medieval folklore credited rosemary and other herbs, such as rose petals, lemon balm, chamomile, costmary, hyssop, meadowsweet, lavender buds, violet petals, mint leaves, heliotrope, lovage leaves, and sweet flag, with other properties of special interest to a bride. As a favorite herb in kitchen gardens, a thriving plant became associated with the strong influence of the mistress of the house on domestic matters. *The Treasury of Botany,* a botanical guide published in the early 1900s, noted a prejudice against rosemary stemming from the folk belief popular in Gloucestershire, England, that it grew best in female-dominated households (Templeton, 2000). "So touchy are some of the lords upon this point, that we have more than once had reason to suspect them of privately injuring a growing rosemary in order to destroy this evidence of their want of authority," the book reported.

Elizabethan Wedding Herbs

Many Elizabethan wedding customs can be found in epithalamums, or marriage odes, that were written to honor a particular bridal couple. These lushly romantic odes provide a glimpse into general herbal lore and wedding traditions of the English renaissance. For the Elizabethan, flowers and herbs symbolized marital love in the full bloom of youth and foretold the expected fruits of their union, linking marriage with the rhythms of nature. The use of the herb marjoram signified marital joy and happiness, an Elizabethan herbal tradition that revealed classical roots. Poets often evoked Hymen, the Roman god of the marriage feast, asking him to bless the bridal couple and bring happiness to their ceremony.

In *A Midsummer Night's Dream,* Shakespeare cleverly inverted poetic conventions that equated a marriage union's harmony with nature's bounty. Fairy queen Titania describes the discord in her relations with Oberon in images of natural order gone awry. "The seasons alter: hoary headed frosts/Fall in the fresh lap of the crimson rose; And on old Hiems' thin and icy crown/An odorous chaplet of sweet summer buds/Is, as in mockery, set" (Act II, Scene I, Lines 107-111). Oberon turns to herbal remedies to set things right.

Rosemary was a popular Elizabethan wedding herb in bridal bouquets. In "The Bride's 'Good Morrow,' " an anonymous Elizabethan poet described a wedding in which "Young men and maids do ready stand/With sweet rosemary in their hand." In 1540, Anne of Cleves wore a bridal wreath of rosemary branches dipped in scented water at her wedding to King Henry VIII of England as his not-so-unlucky fourth wife. Although the marriage did not last long, she was amicably divorced from the monarch and lived in much comfort and contentment until her death in England.

Victorian Wedding Herbs

By the nineteenth century, marriage had become a social display as Americans and Europeans replaced the simplicity of the past with distinctly Victorian innovations, such as a white dress for the bride, color themes for decorations and bridal party gowns, wedding attendants that included flower girls and ring bearers, wedding favors for the guests, formal reception lines, and a scent-filled, glorious bridal bouquet—all harbingers of the billion-dollar wedding industry that was to follow. Early in the century, brides carried small bouquets, arranged as tussie mussies, and a prayer book. Midcentury bouquets contained more formal displays of white roses along with trailing ribbons to add floral accents and intricate knotwork. By 1900, at least two dozen roses were required for a proper bridal bouquet.

For her marriage in February 1840, Queen Victoria carried a bouquet decorated with myrtle greens, a symbol of home and love since ancient Rome. After the ceremony, the farsighted monarch had myrtle from the bouquet planted in the royal garden. Since then, subsequent British royal brides have carried a sprig or two of myrtle from one of Victoria's long-living plants in their wedding attire.

Like their Elizabethan counterparts with their floral symbolism, Victorian brides communicated through the language of flowers but translated the Elizabethan wedding revelry into a more focused display of youth and innocence. As an example, marjoram, that ancient emblem of Hymen and Aphrodite, symbolized the joy of married life to Shakespeare and his contemporaries; however, in *Gems of Deportment, A Manual Instruction for the Home* (1881), marjoram represented "blushes" (modesty). In the Victorian tradition of the language of flowers, rosebuds symbolized undying love; lavender, devotion; rosemary, remembrance; orange blossoms, bridal festivities; marjoram, joy; vanilla beans, good luck; sage, domestic virtue; oakmoss, ardent love; bergamot, compassion; and frankincense, a faithful heart.

The legacy of wedding herbs continues as each generation borrows past traditions and creates new customs as well. Herbal accents evoke medieval floor rushes, the classicism of Renaissance plant rituals, and the sentimental sweetness of Victorian floral confections. Native herbs and plants in wedding decorations symbolize community, including herbal designs in wedding invitations, herbal seeds as wedding favors, and sachets of personal favorites that continue to provide wedding apparel with its own distinctive scent.

BIBLIOGRAPHY

Shaudys, Phyllis V. *Herbal Treasures: Inspiring Month-By-Month Projects for Gardening, Cooking, and Crafts*. Pownal, VT: Storey Communications, 1990.

Templeton, Lora. "Something Old, Something New." *Herb Quarterly 85* (Spring 2000): 29-35.

HOPS

Ancient legend relates that the Egyptian sun god, Ra, first taught man how to brew beer. In the first century B.C., Diodorus Siculus observed that the German tribes drank beer seasoned with hop *(Humulus lupulus),* a practice derived from Finnish tribes to the north. The shoots of the plant were cooked like asparagus by the Romans, and served in salads with a vinaigrette dressing. By the eighth century, Benedictine monks included hops in their herb gardens. Arabian doctor Mesue noted in the midninth century that hops could induce sleep and improve the nervous system.

Its value in enhancing flavor and as a preserving agent caused most European countries to include hops in the brewing process during the Middle Ages. The practice was evidently widespread in France prior to the 1500s; from there it was introduced into the British Isles for cultivation purposes (although it had been growing wild there for centuries). Its use in England was a matter of controversy for several hundred years. Critics noted the groggy feeling resulting from its ingestion. In 1670, John Evelyn would observe, "Hops . . . preserves the drink, but repays the pleasure in tormenting diseases and a shorter life." Growing the herb was forbidden under the reign of Henry VI (1422-1461), but Henry VIII (1509-1547) allowed a limited degree of cultivation, albeit not in the brewing of beer. By the seventeenth century, however, hops were permitted in British beermaking for good. The process was introduced into the American colonies at that time as well.

The majority of the world's supply presently comes from the United States (the far western states), Germany, and Great Britain; Europe, Australia, New Zealand, Japan, and Argentina also cultivate hops for use as a beer-flavoring agent. Hops is also used to flavor tobacco, yeast, beverages, confections, baked goods, chewing gums, condiments, and as an aromatic bitter principle in pharmacy.

Perhaps less widely known to the public is the fact that hops possess a long history of use as a medicinal. The portion of the plant employed for this purpose, the female flower (or strobile), has a bitter, slightly pungent taste. The strobile contains up to 1 percent volatile oil, which in turn includes over 100 compounds, most notably bitter resin (three to twelve percent; composed of lupulone, lumulone, and valerianic acid), lumulene, mycrene, beta-caryophyllene and farnescene, geraniol, linalool, citral, linionene, and serlidol in addition to tannins, fats, amino acids, phytoestrogens, and flavonoids. Fresh strobiles facilitate optimum activity; with aging, they tend to stimulate rather than calm the user.

The beneficial physiological effects of hops include:

- Reducing pain
- Easing tension (by taking tea or some other form of tincture)
- Inducing sleep (through the use of pillows filled with fresh strobiles)
- Stimulating the digestive tract (via the ingestion of capsules)
- Relaxing smooth muscle, thereby (in combination with herbs such as valerian and skullcap) relieving the symptoms of Crohn's disease, irritable bowel syndrome (IBS), and nervous stomach
- Acting as a diuretic
- Producing an antibacterial effect
- Functioning as an anti-inflammatory, particularly in the reduction of gastric and duodenal ulcers

Hops require caution in handling as they can cause a number of adverse reactions. When absorbed through the skin, the estrogenic compounds within their volatile oil can interrupt the menstrual cycle and decrease male libido. Likewise, an abundance of these compounds can complicate pregnancy. Keeping skin covered when picking the flowers is advised because the plant's pollen can cause dermatitis.

Anyone whose symptomology or constitution is aggravated by the cold and drying nature of hops should discontinue use. Due to the depressant effect on the human nervous system, hops should also be used sparingly in cases of depression. Furthermore, hops "spent" in the brewing process must be discarded carefully because certain breeds of dogs (e.g., greyhounds) can die from ingesting them.

BIBLIOGRAPHY

Farrell, Kenneth T. "Hops." In: *Spices, Condiments, and Seasonings*. Westport, CT: AVI, 1985, pp. 127-129.

Langer, Robin E. "Herb-of-the-Month: Hops *(Humulus lupulus)*." *Snowbound Herbals* (August 1999): 2p. Updated September 6, 2000 at Web site www.sbherbals.com.

Moyler, D.A. "Hop Cones." In: Charalambous, George (Ed.), *Spices, Herbs and Edible Fungi*. Amsterdam: Holland, 1994, pp. 32-35.

HORSETAIL

Although reaching no more than eight to ten inches in height, horsetail *(Equisetum arvense)* is a descendant of carboniferous era plant that grew as tall as trees. Sometimes called shavegrass, mare's tail, pewterwort, scouring rush, or bottlebrush, its distinctive appearance consists of a single stem punctuated by bristlelike branches growing out of the joints. A small, brown, cone-shaped flower appears on top in early spring. Leaves appear after the flowers have died. Like ferns, spores from the flower are the plant's mechanism for propagation. It spreads rapidly and is frequently difficult to eradicate.

Horsetail's medicinal value results from its mineral and silica content. The latter helps strengthen many body tissues, including bone, hair, and nails as well as promoting calcium absorption and combating plaque deposits in the arteries. The presence of minerals such as manganese, potassium, sulfur, and magnesium can treat anemia and general debility.

Horsetail has a number of other health-related uses, many of which stem from its astringency and diuretic action. These include:

- Soothing inflammation (e.g., cystitis) and treating infections within the urinary tract in addition to reducing urinary frequency and incontinence
- Fighting prostate problems
- Hastening the removal of kidney stones
- Assisting in the control of bed-wetting
- Treating comprehensive lung damage (e.g., tuberculosis, emphysema)

- Stopping bleeding from wounds (by applying bruised, fresh horsetail) and ulcers or heavy menstrual bleeding (the latter two by drinking juice or tea)
- Easing sore throat, bleeding gums, and mouth ulcers (by gargling and rinsing the mouth with tea)
- Strengthening brittle hair (via a hair rinse)
- Mitigating irritated skin conditions such as eczema, scrapes, and ulcers (by means of a poultice)
- Treating fungus-infected toe- or fingernails (through soaking in horsetail vinegar)

Furthermore, the dried stems of the plant have been effective over the centuries as an abrasive. Prior to the appearance of modern-day chemical products, horsetail was widely used to scour pots and polish wood, brass, and pewter.

Because it interferes with thiamin absorption, anyone taking the herb should supplement it with vitamin B-1.

BIBLIOGRAPHY

Hollis, Sarah. "Horsetail." In: *The Country Diary Herbal.* New York: Henry Holt, 1990, pp. 60-61.

Langer, Robin E. "Herb-of-the-Month: Horsetail *(Equisetum arvense)*—May 1998." *Snowbound Herbals* (May 1998): 2p. www.sbherbals.com.

HORTICULTURE THERAPY

Betty Stevens, a Connecticut-based herb grower, states that horticulture therapy became a recognized profession in the early 1980s. She has outlined possible areas of specialization within this newly emerging field:

I. Landscape design and garden construction
   A. Herb gardens in planters and beds at wheelchair level for convalescent homes and rehabilitation centers
   B. Culinary herb gardens for senior centers, convalescent hospitals, rehabilitation centers

1. Introduction of herbs into cooking, especially helpful in various diets
2. Drying and freezing for winter use

C. Mixed gardens for tending and using in cooking and craft work in all types of institutions

II. Programs within the facility

A. Bedside gardening for confined clients and terminally ill: tussie mussies, seeds, dish gardens

B. Craft projects: small wreaths, tussie mussies, potpourri and sachets (p. 7)

The benefits of such programs would include the promotion of well-being in institutionalized clients and aesthetic improvement of the environment. The medical sciences literature provides many examples of the positive impact of horticulture therapy.

Negative features include the short-term character of many of the projects and the lack of funding in many health care institutions for anything other than basic services. The continued expansion of the profession—for example, the development of new jobs such as that of herb consultant—should address the former concern. Funding limitations can be overcome by aggressive grant-seeking or the willingness to implement projects on a volunteer basis. Stevens argues that funding sometimes comes once a program has been established and the benefits are clearly seen.

BIBLIOGRAPHY

Stevens, Betty. "Horticulture Therapy." *The Business of Herbs 1*(6) (January/February 1984): 7.

HYSSOP

A bushy evergreen reaching one to two feet in height, hyssop *(Hyssopus officinalis)* thrives in dry, rocky, calcareous soil receiving full sun. The small, narrow leaves grow in whorls on the squarish stem, much like the flowers. The latter appear from June to October and are dark blue in color; some varieties, however, are pink, purple, or white. The plant is propagated by seed or cuttings taken from late spring to early summer. Its aerial parts (those portions growing above ground) are harvested for medicinal use immediately before or during the flowering stage.

A native of southern Europe, hyssop was spread as far north as the British Isles by the Romans during antiquity. It went on to become a fixture in monastery infirmary gardens. It was once held as a symbol of purification and was employed in the ritual cleansings of churches and leprous houses.

The active ingredients of hyssop include up to 2 percent volatile oils (primarily pinocamphone, isopinocamphone, pinenes, camphene, and terpenine), a glycoside known as hyssopin, tannins, flavonoids, isolic acid, olenolic acid, marrubin (largely responsible for its pungent, bitter taste), resin, gum, and approximately fifty other compounds. Its expectorant and diaphoretic features render it effective in relieving symptoms of congestion from chest colds, bronchitis, and influenza. The plant has diuretic properties, relieves flatulence or an upset stomach, and can remove worms. Topically applied, it treats scalding, minor burns, and bruises. It can be taken as a tincture (alcoholic extract), a syrup, a tea, or an essential oil.

Hyssop is one of the main herbs used to flavor liqueurs such as Benedictine and Chartreuse. It possesses an exceptionally strong flavor; although effectively employed to disguise the taste of rancid meat in the past, it is now viewed as too overpowering for general culinary use. The plant's strong scent made it a valuable component in nosegays, and its volatile oil has long been a favorite with perfumers. Because hyssop retains its scent when dried, it is widely used in potpourri and herb pillows.

BIBLIOGRAPHY

Hollis, Sara. "Hyssop." In: *The Country Diary Herbal.* New York: Henry Holt, 1990, pp. 62-64.

Langer, Robin E. "Herb-of-the-Month: Hyssop *(Hyssopus officinalis)*—March 1997." *Snowbound Herbals* (March 1997): 2p. www.sbherbals.com.

I

INDIANA BOTANICAL GARDENS

Indiana Botanical Gardens, the nation's oldest and largest retailer of herbs, was founded in 1910 by Joseph Meyer. IBC focuses on milling, packaging, and reselling herbs, teas, and vitamins to physicians, health stores, and the general public. The store is located at 3401 W. 37th Avenue, Hobart, Indiana; however, the bulk of the business has been handled through mail order for many decades. Their Web site is www.botanichealth.com.

The retailer makes many of its own teas, shampoos, and other herbal products. It is licensed by the Food and Drug Administration as a manufacturer of drugs and is required by that agency to undergo inspections twice a year. IBC executive David Meyer noted in 1984, "Our standards are very high. Overall we reject about 30 percent of the botanicals that come in for processing." Reasons for rejection include being too old, moldy, or contaminated, or having been picked at the wrong time of year.

IBC has exhibited a willingness to educate and encourage entrepreneurs interested in growing herbs. Meyer has indicated that most herb-based ventures don't appear to be long-lasting. He attributes this high failure rate to a lack of experience: "They find out it's too much work for the return they get. Or they simply don't know enough about the plant they are trying to grow."

BIBLIOGRAPHY

Meares, Portia. "Indiana Botanical Gardens: Selling Botanicals for Four Generations." *The Business of Herbs 1*(6) (January/February 1984): 1, 3, 9.

INTERNATIONAL HERB ASSOCIATION

The impetus for the formation of the International Herb Association (IHA) was a 1985 meeting of businessmen in Lebanon, Ohio, organized to discuss common ideas and goals. Organizers expected some twenty-five businesses from Ohio and Kentucky to attend; to their surprise, ninety-two people representing fifty-six businesses took part. A steering committee was formed; at Purdue University the following year, the International Herb Growers and Marketers Association was organized (renamed the International Herb Association in 1994). Since 1986, educational conferences and business meetings have been held on an annual basis with the aim of "uniting herb professionals."

From its inception, IHA has encouraged herbal entrepreneurs as an association of peers supporting herbal endeavors. IHA provides the latest information and technology on growing, marketing, and using herbs and involves its membership in the decision-making process. The organization is governed by a nine-member Board of Directors elected by the membership for three-year terms. Three new members are elected each year. The International Office implements the Board's decisions, handles day-to-day concerns, supplies information about the organization, and maintains its records. In 1996, the first regional chapter, the Southeast, was formed. Other regional chapters and affiliations are under consideration.

IHA membership brings many benefits, among them access to people and resources throughout the herb industry, including a speakers' bureau. While centered in North America, the membership includes many other nations in a growing worldwide family. IHA members represent virtually every profession, a virtual "Who's Who" of Herbs. Another benefit includes Web pages for members at significant savings. Members can maintain their own Web sites without owning a personal computer; the IHA is providing guidance in home page development in conjunction with a professional Webmaster. IHA publications include a membership directory and the bimonthly *IHA Newsletter.*

Members tend to be either hardcore herbal enthusiasts or business owners. However, those involved in allied industries or organizations, individuals planning an herb-related business, and anyone interested in serious herb study could profit from joining the IHA. The organization's annual conference addresses diverse concerns of growers, marketers, and researchers. Topics typically covered include packaging and labeling, determining sources of supplies, mail order, computers, display and design, taxes, employee relations, stress management, and business budgets and projections. In 1995, the IHA instituted an Herb of the Year award. Fennel was the first plant to be recognized, followed by monarda or bee balm (1996), thyme (1997), mint (1998), lavender (1999), and rosemary (2000).

The International Herb Association has headquarters at 910 Charles Street, Fredericksburg, VA 22401-5810. Phone: (540) 368-0590; fax: (540) 370-0015; e-mail: info@iherb.org; Web site: www.iherb.org.

BIBLIOGRAPHY

International Herb Association Home Page: www.iherb.org.
International Herb Association. *IHA Newsletter* [Various issues, 1995-2000].

INTERNATIONAL HERB GROWERS AND MARKETERS ASSOCIATION

(*See also:* INTERNATIONAL HERB ASSOCIATION)

The International Herb Growers and Marketers Association (IHGMA) was instituted in 1986 to promote the use of herbs and provide support for herb growers. Past president Dick Tippett notes, "The herb industry is wide ranging, from the big conglomerates—the big tea and spice companies—to the individual crafters." IHGMA's target constituency encompasses those who sell fresh-cut herbs to restaurants, grow medicinal herbs, and grow and manufacture decorative herbs. According to another former president, Portia Meares, the latter category includes "the whole craft scene—herb products, potted herbs and posters."

The formation of the group was, in large part, a response to rapidly increasing herb sales during the early 1980s. U.S. Department of Agriculture statistics for 1987 revealed that herb sales rose 63 percent during the five previous years; 28.8 million pounds were sold in twenty-three major U.S. markets.

Activities of the IHGMA include sponsoring research and development, publishing a newsletter, coordinating an annual conference, and educating the public about the benefits of herbs. In the early 1990s, the organization applied to the U.S. Postal Service to feature herbs on stamp issues. The postal service's decision to produce a series of herb stamps enabled wholesalers and retailers to employ production, advertising, and marketing strategies to greater effect.

In 1990, the IHGMA engaged Stygar Associates to serve as executive director. The Mundelein, Illinois–based management firm assumed its administrative duties on December 1, 1990. As a result, the IHGMA, then over 1,000 members strong, entered the new decade with a renewed sense of purpose and professionalism. In order to reflect its broader slate of concerns, the organization changed its name to the International Herb Association in 1994.

BIBLIOGRAPHY

"Stygar Associates To Manage Herb Group." *Nursery Manager* (December 1990): 12ff.

J

JUNIPER

Found in abundance across Europe, North Africa, Asia, and North America, juniper *(Juniperus communis)* is a perennial whose distinctive features include thin, pointed, gray-green leaves and a distinctive fragrance. The plant grows close to the ground when exposed to mountain winds, while achieving ten feet or more in height in protected settings. Small cones—males and females on different plants—appear at the base of the leaves in early summer. The male cones are yellow; the female variety—bluish-green in color—grow into fleshy berries, taking on a blue-black hue while ripening over a three-year span.

Juniper has been employed medicinally as far back as 1550 B.C. when papyrus documents from Egypt mentioned it as a tapeworm remedy. Furthermore, branches and berries were burned at the time in purification ceremonies. The Ayurvedic healing system of India holds that the plant purifies both the physical body and the aura (or subtle body) in addition to destroying negative astral influences. Juniper oil was used to treat many conditions, including typhoid, in the Middle Ages.

Today, the berries are in demand to flavor gin, as a culinary accent, and in herbal medicine. They are diuretic, diaphoretic, stimulant, carminative, analgesic, disinfectant, bactericidal, and antirheumatic, as well as being a uterine stimulant, antifungal, emmenagogue, and insect repellant. Juniper can be used medicinally in a number of forms: cade oil or juniper tar oil (derived from dry distillation of the heartwood, it treats chronic skin conditions); essential oil (steam distillation of the berries); teas or tinctures; ingesting the berries, especially in soups or stews; and ointments or lotions (berries soaked in olive or other pure vegetable oil). Conditions it can treat include eczema, psoriasis, and other skin problems as well as cystitis, urethritis, weak digestion, acid waste buildup, rheumatism, gout, sore muscles, menstrual cramps, and coughs.

The plant does have its caveats. Since it can be employed during labor to improve uterine contractions, women are advised to avoid it during pregnancy. Likewise, the very young and senior citizens are considered at-risk users.

BIBLIOGRAPHY

Griggs, Barbara. *Green Pharmacy: The History and Evolution of Western Herbal Medicine.* Rochester, VT: Healing Arts, 1997.

Langer, Robin E. "Herb-of-the-Month: Juniper (Juniper communis)." *Snowbound Herbals* (September 1999): 2p. Revised September 6, 2000. www.sbherbals.com.

Moyler, D.A. "Juniper (berry)." In: Charalambous, George (Ed.), *Spices, Herbs and Edible Fungi.* Amsterdam: Elsevier, 1994, pp. 60-61.

K

KAVA KAVA

Kava kava *(Piper methysticum)* has enjoyed considerable popularity in recent years for combating anxiety. As of mid-2001 it was included in "Top Ten Selling Herbs," a list compiled by a leading Internet retailer, Discount Natural Herbs. Kava kava's success is all the more impressive considering that few herb guides even bothered to profile it prior to the late 1990s.

Medical studies have indicated that the herb is not only as effective as prescription tranquilizers, it is nonaddictive and does not impair mental function or coordination at the recommended dosages (i.e., forty-five to seventy milligrams kavalactones, up to three times daily). In many cases, individuals who initially tried mainstream pharmaceuticals reported increased alertness and improved concentration after switching to kava.

A warning must be heeded, however, with regard to the use of kava. High doses of kava root extract are known to be intoxicating, resulting in sedation and extreme muscle relaxation. Ceremonial rituals among certain native cultures of the South Pacific—which have institutionalized such high-intake levels—provide ample documentation regarding these symptoms.

Furthermore, the fact that its active ingredients, the kavalactones, appear to affect the brain biochemically in much the same manner as potent prescription tranquilizers (e.g., benzodiazepine), has led to a rash of sensationalized media accounts concerning kava use. Dr. Robert Rountree, a Boulder, Colorado–based, physician who has advocated the use of kava, states that misleading information has unfairly stigmatized the herb. In his article, "The Safety of Kava Kava," Rountree focuses on a high-profile story first published in 1996 as a letter to the editor of a prominent medical journal and then featured in media outlets nationwide under the following heading: "Coma from the Health Food Store: Interaction between Kava and Alprazolam." He notes that, on closer inspection, a completely different story emerged out of the incident:

> A fifty-four-year-old man was hospitalized in a *"semi-comatose"* state, which the authors described as "lethargic and disoriented." He was never in a coma—the term "semi-comatose" appears to have been an exaggeration. He began to improve within a few hours, apparently without any specific medical intervention.

He added that the patient was found to have been taking three prescription drugs at the time: cimetidine, an antacid; alprazolam, a tranquilizer known to cause drowsiness and fatigue; and terazosin, a blood pressure medicine with side effects that included weakness, lassitude, and fatigue.

> And cimetidine is notorious for its effect on the metabolism of other drugs, such as the benzodiazepine class of tranquilizers that includes alprazolam. By inhibiting the breakdown of those drugs in the liver, it's as if a much higher dose was being taken, thus increasing the risk of side effects.

Three days after beginning to take kava (purchased from a health-food store in an unknown form and dosage), the patient's symptoms developed.

Herb advocates have argued that it defies logic to take kava together with a prescription tranquilizer—or, for that matter, with alcohol, barbiturates, sleeping pills, and psychotropic drugs. In short, the resulting effects are likely to be additive and possibly synergistic. Rountree advises that the public, rather than believing that all herbal medicines are nostrums, should exercise common sense in applying an herb-based treatment plan. For example, herbalists frequently have combined kava with other herbal agents that calm the nervous system (e.g., valerian root, hops, skullcap) without any documented evidence of toxicity.

BIBLIOGRAPHY

Heinerman, John. "Kava Kava *(Piper methysticum).*" In: *Heinerman's Encyclopedia of Healing Herbs & Spices.* Paramus, NJ: Prentice-Hall, 1996, p. 296.

Rountree, Robert. "In Defense of Herbs." *Herbs for Health* (January/February 2000). 1p. Includes hyperlinks to the following articles: "The Rundown on Kava"; "The Safety of Kava Kava." www.discoverherbs.com/features/kava_indefense.html.

Rountree, Robert. "The Safety of Kava Kava; Don't be Mislead by Scare Tactics." Interweave Press, 1999. 2p. www.discoverherbs.com/features/kavakava.html.

KUDZU VINE

A member of the pea family (Leguminosae), the kudzu vine *(Pueraria lobata)* was imported into the United States from Japan prior to 1876 to counter soil erosion. Demonstrating promise as a food and fodder plant, it came to be viewed as an invasive pest by the end of World War II, covering a half million acres in the southeastern states.

Pueraria species have long been valued in China and India as aphrodisiacs, contraceptives, diuretics, and treatments for angina pectoris and high blood pressure. In the late 1990s, Jiwaji University (India) researchers determined that a root abstract of *Pueraria tuberosa* stabilized the activity of liver enzymes, stimulated regeneration of liver tissue, and made the liver more resistant to damage from carbon tetrachloride toxins.

BIBLIOGRAPHY

Foster, Steven. "Plants for Pain Relief." *The Herb Companion* (August/September 1997): 60-63.

"New Uses for an Old Weed." *The Herb Companion* (October/November 1996): 60.

Shukla, S. et al. "Protective Action of Butanolic Extract of *Pueraria tuberosa* DC. Against Carbon Tetrachloride-Induced Hepatotoxicity in Adult Rats." *Phytotherapy Research 10*(7) (1996): 608-609.

L

LAVENDER

Lavender *(Lavandula officinalis)* is native to southern Europe; however, it is widely cultivated around the world, particularly in dry, sunny locales. A woody-stemmed shrub reaching six to twenty-four inches in height, its leaves are opposite, narrow, three-fourths to two inches in length, and gray-green in color. Its flowers generally appear from June to September; they are small, baby-blue in color, strong-smelling, and culminate in spikes at the end of long stalks. Due to changes in nomenclature over the years, misnaming of lavenders continues to plague the marketplace; *Lavandula officinalis* is now known as *Lavandula angustifolia* (English lavender or true lavender).

Lavender was extremely popular during antiquity. The Greeks utilized it as a healing herb as well as for making wines and vinegars. The Romans used it as a scent for bathing and to relieve aching limbs. During the Middle Ages, lavender thrived in monastery gardens.

Herbalists documented the herb's value as a laxative and tonic and claimed it could heal anything from colds and headaches to limb paralysis and neurosis. It was one of the main ingredients of "Four Thieves Vinegar" used to fight the bubonic plague in the fourteenth century. In addition to its diversified uses as an ornamental plant, in cosmetics, and as a food flavoring, lavender possesses many medicinal applications. It has been used to treat minor burns; insect bites; emotional problems such as anxiety, depression, and stress; coughs; headaches; and infantile colic. Fresh spikes of lavender are suspended in water in closed bottles exposed to direct sunlight, to be utilized for cleansing wounds. Furthermore, scientific research in the mid-1990s determined that its smell stimulates sexual arousal in males.

BIBLIOGRAPHY

Gardner, Jo Ann. "English Lavender." In: *Living with Herbs*. Woodstock, VT: Countryman Press, 1997, pp. 178-181.

Heinerman, John. "Lavender *(Lavandula officinalis)*." In: *Heinerman's Encyclopedia of Healing Herbs & Spices*. Paramus, NJ: Prentice-Hall, 1996, pp. 299-304.

LEMON BALM

Lemon balm *(Melissa officinalis)* is a hardy perennial possessing decorative, cosmetic, culinary, aromatic, and medical uses. The plant is easily propagated by division, cutting, or seed. While it will grow in dry soils, the scent is stronger when the ground is moist and fertile. Lemon balm matures into a rounded mound two to four feet in height; the leaves resemble those of mint. Its small white flowers, which cluster up the stem in summer, are extremely attractive to bees. Cultivated plants are often replaced by gardeners every two or three years because the old plants become overlarge and lose the neat form of youth.

Herbalists typically place lemon balm next to a path or in a low raised bed, where its pronounced lemon scent is released as the leaves are brushed in passing. The "All Gold" and "Variegata" varieties (the latter of which tolerates full sun) are preferred in the front of shrubs or herbaceous borders.

Lemon balm is widely used as a seasoning; for example, it can be used in salads and other vegetable dishes, milk custards, and as a substitute for lemon with whitefish. It is often used as an herbal tisane, or tea, for the treatment of headache, insomnia, nervous tension, and depression.

The preferred mode for medicinal applications of lemon balm—as with a tisane—is generally use of the aerial parts in infusion or tincture. However, the herb is also employed to treat insect bites and stings through application of fresh leaves.

The lemon balm leaf has a variety of cosmetic uses as well. An infusion is supposed to promote relaxing baths. In addition, a tincture or infusion can produce a scented water cologne.

In the days prior to the proliferation of commercial polishes and cleaners, the herb played an important role in various household preparations. The leaves were traditionally added to furniture polish as well as infused in linseed oil and turpentine. Lemon balm was also a fixture in laundry rinses. It continues to be widely used as an air freshener in potpourri mixes. The extent of lemon balm's value to present-day consumers is reflected by its inclusion in Deborah C. Harding's *The Green Guide to Herb Gardening: Featuring the 10 Most Popular Herbs* (2000).

BIBLIOGRAPHY

Gardner, Jo Ann. *Living with Herbs*. Woodstock, VT: Countryman Press, 1997.

Garland, Sarah. *The Complete Book of Herbs & Spices*. Pleasantville, NY: Reader's Digest Association, 1993.

Harding, Deborah C. *The Green Guide to Herb Gardening: Featuring the 10 Most Popular Herbs*. St. Paul, MN: Llewellyn, 2000.

Leigh, Evelyn. "Valerian/Lemon Balm Combination Well Tolerated in Clinical Study." *HerbalGram 49* (2000): 27.

McHoy, Peter, and Westland, Pamela. *The Herb Bible*. New York: Barnes & Noble Books, 1994.

LEMONGRASS

Notable for its pronounced citrus flavor, lemongrass *(Cymbopogon citratus)* is native to the tropical regions of Southeast Asia. It is cultivated today in Africa, Australia, India, South America, and the United States.

The key ingredient in the leaves is citral. It is extremely popular as a fragrance component in creams, detergents, lotions, perfumes, and soaps. Lemongrass oil is a component in most major food groups, including alcoholic and nonalcoholic beverages, baked goods, confections, fats and oils, frozen dairy desserts, gelatins and puddings, fish and meat dishes, and soups.

Although less widely known, the herb has proven effective as a pain reliever. A tea brewed from its leaves helps combat fevers. Taken orally, lemongrass oil provides relief from colic, intestinal gas, and obstinate vomiting. Rubbed into the skin, it relieves lower backache, lumbago, rheumatism, sciatica, sprains, and tendinitis.

A related species, citronella *(Cymbopogon nardus),* is also widely used as a fragrance component and food additive. Widely cultivated in Africa, Indonesia, Latin America, Malaysia, and Sri Lanka, the oil reportedly possesses antibacterial and antifungal activities in vitro. In addition, it can serve as an insect repellent. Johnson Wax's popular formulation, Off!, is based on citronella oil.

BIBLIOGRAPHY

Heinerman, John. "Lemongrass *(Cymbopogon citratus).*" In: *Heinerman's Encyclopedia of Healing Herbs & Spices.* Paramus, NJ: Prentice-Hall, 1996, pp. 305-306.

McHoy, Peter, and Westland, Pamela. *The Herb Bible.* New York: Barnes & Noble Books, 1994.

LICORICE

Licorice *(Glycyrrhiza glabra)* is native to dry scrubland and damp ravines in both the Mediterranean region and southwestern Asia. It was brought to Great Britain in the late Middle Ages by Dominican friars; today, it is cultivated worldwide, including in Russia, Greece, Spain, the Middle East, Italy, India, Arizona, and California.

The plant reaches three to five feet in height and possesses pinnately compound leaves of nine to seventeen roughly one-inch long elliptical leaflets. In late summer, short, stalked spikes of blue or violet flowers bloom in the leaf axils. The flowers, approximately one-half inch in length, are followed by smooth, oblong pods one-half to one inch long containing two to four seeds.

The medical applications of licorice have been documented since antiquity. Greek philosopher and naturalist Theophrastus (372-287 B.C.) reported that licorice root was used to treat respiratory disorders and, when ingested with mare milk cheese, enabled the Scythians to go as long as twelve days without water. The Chinese have employed the root as an antidote to a wide range of illnesses; believing that it reduces the toxicity of certain substances, they still include licorice in many formulations. Native Americans and early European settlers utilized the related plant, *Glycyrrhiza lepidota,* to induce menstrual periods, expel the placenta after childbirth, and relieve earache, toothache, and fever. In addition, licorice has been applied to the treatment of sore throat, urinary tract infections, stomach ulcers, constipation, Addison's disease, and, externally, to soothe irritated skin and eyes.

In the August/September 1997 issue of *The Herb Companion,* Betsy Strauch notes that scientific studies underscore the value of the plant in many of these applications. Various research findings have indicated that licorice (or derivatives of the root) can

- Relieve inflammation associated with arthritis, skin irritations, and respiratory conditions
- Protect the liver and keep red blood cells from rupturing in the presence of certain toxins

- Help prevent adrenal gland atrophy in patients receiving conventional drug treatment for Addison's disease
- Be effective as hydrocortisone without the latter's side effects
- Spur the production of interferon, a key chemical in the immune system
- Inhibit the growth of bacteria, including drug-resistant strains, and viruses
- Stimulate normal ovulation in women with infrequent menstrual periods
- Act as a mild laxative (taken with stronger laxatives, it makes them less irritating to the digestive tract)
- Suppress coughs as effectively as codeine
- Heal ulcers of the stomach and small intestine (when glycyrrhizin, the cortisonelike component responsible for licorice's sweetness, has been removed from the herb)

Licorice also has also found a wide variety of nonmedicinal applications. In recent years it has been used in

- Beer and liqueurs, ice cream, laxatives, mouthwash, soft drinks, soy products, and toothpaste to enhance flavor
- Beers and fire extinguishers as a foaming agent
- Insecticides as a spreading agent
- Fertilizer (like other legumes, it possesses the ability to fix nitrogen from the air)
- Chocolate as a sugar extender
- Cosmetics (as a result of its skin-soothing qualities)
- Composition board and insulation

Because the long-term use of the plant can result in sodium retention and potassium loss, pregnant women, individuals with high blood pressure or kidney disease, and those taking digitalis medications are advised not to ingest licorice. However, the author of *The Scientific Validation of Herbal Medicine,* Daniel B. Mowrey, has maintained that whole licorice roots (or products made from them) are not toxic. Rather, virtually all recorded cases of licorice toxicity are the result of taking concentrated root extracts or derivatives (e.g., carbenoxolone sodium, an ingredient in a number of ulcer remedies).

BIBLIOGRAPHY

Foster, Steven. "Licorice: Sweet Medicine." *The Herb Companion* (October/November 1996): 58-59.

Garland, Sarah. *The Complete Book of Herbs & Spices.* Pleasantville, NY: Reader's Digest Association, 1993.

Mowrey, Daniel B. *The Scientific Validation of Herbal Medicine.* New Canaan, CT: Keats Pub., 1986.

Strauch, Betsy. "An Herb To Know." *The Herb Companion* (August/September 1997): 20-21.

LOBELIA

An annual or biennial that reaches approximately a yard in height, lobelia *(Lobelia inflata)* is found in the meadows and cultivated fields of the Midwest, eastern United States, the Rocky Mountain belt, and Pacific Northwest. It possesses an erect, angular stem and exudes a milky sap when broken or cut. The leaves are light green, alternate, ovate, bluntly serrate, and (like the rest of the plant) hairy. A profusion of tiny, two-lipped, blue flowers grow on top in spiky racemes between July and December.

In the mid-nineteenth century, lobelia was central to the eclectic botanical healing system originated by Samuel Thomson, known as Thomsonian Medicine. It was once widely utilized for respiratory and inflammatory disorders as well as to eliminate the rigidity of pelvic muscles during childbirth. The herb's use today is limited largely to herbal medicine practitioners who consider it effective as an anti-inflammatory in cases of erysipelas, as an antispasmodic for whooping cough, as a smoking deterrent, and for diphtheria and tonsillitis.

The herb is ingested either as a tincture or tea. The former is a stimulating concoction to say the least; it consists of three-fourths cup of cut fresh or powdered lobelia mixed with one pint of English brandy, Jamaican rum, Russian vodka, or Southern gin. The mixture is shaken twice daily for sixteen days. After settling, it should be strained through cotton or a paper coffee filter during a full or new moon to assure maximum potency (author John Heinerman asserts this recipe has a scientific basis). Droplets of the tincture are either taken beneath the tongue or mixed with distilled water for treating a variety of condititions.

BIBLIOGRAPHY

Griggs, Barbara. *Green Pharmacy: The History and Evolution of Western Herbal Medicine*. Rochester, VT: Healing Arts, 1997.

Heinerman, John. "Lobelia *(Lobelia inflata)*." In: *Heinerman's Encyclopedia of Healing Herbs & Spices*. Paramus, NJ: Prentice-Hall, 1996, pp. 313-314.

Teich, Mark, and Weintraub, Pamela. "Are Natural Cures a Prescription for Danger?" *Redbook* (June 1995): 88-89, 111-114.

M

MAHUANG

A small- to medium-sized shrub, mahuang *(Ephedra sinica)* looks much like its relative, Brigham Tea. The barkless stems are jointed with spikelike tips; they vary in color from grayish-blue to bright yellow-green. The plant produces degenerate leaves at the nodes that are reduced to scales, either two or three depending on the species.

Mahuang was extremely popular up until the late 1990s as the main ingredient in many over-the-counter drugs and herbal supplements used for energy and weight loss. It is the source of the natural constituents ephedrine and pseudoephedrine, used to treat upper respiratory ailments. The latter variant of ephedrine was the chief ingredient in Contac, Robitussin-PE, Sinutab, Sudafed, Actifed, and other bestselling pharmaceuticals.

In the face of increasing evidence regarding the risks surrounding the use of the herb, the Food and Drug Administration issued public warnings that overconsumption of herbal energy stimulants containing both mahuang and kola nut or guarana could induce a heart attack and damage the nervous system. Furthermore, physicians recommended avoidance of mahuang by individuals with general weaknesses, the tendency to perspire, poor digestion, hypertension, nervousness, insomnia, cardiac arrhythmias, and heart disease.

By 1995, a number of states had begun legal maneuvering aimed at restricting the sale of natural food supplements and drugs derived from mahuang or ephedrine. According to the January 1995 issue of *Health Foods Business* (p. 6), the Ohio State Board of Pharmacy enacted regulations to restrict sales of ephedrine to pharmacists (cited in Heinerman, 1996, p. 320). The *Fort Worth Star-Telegram* (March 20, 1995, Section A, p. 14) reported that the Texas Board of Health moved to prohibit the sale of food and dietary products containing ephedrine, except in naturally occurring amounts (e.g., mahuang or Brigham Tea) (cited in Heinerman, 1996, pp. 319-320).

BIBLIOGRAPHY

Heinerman, John. "Mahuang *(Ephedra sinica)*." In: *Heinerman's Encyclopedia of Healing Herbs & Spices*. Paramus, NJ: Prentice-Hall, 1996, pp. 319-320.

Teich, Mark, and Weintraub, Pamela. "Are Natural Cures a Prescription for Danger?" *Redbook* (June 1995): 88-89, 111-114.

MARIJUANA

Marijuana (or, more specifically, its primary active ingredient, tetrahydrocannabinol, or THC) is the drug derived from the flowering tops, stems, and leaves of the hemp plant. It has been employed to achieve euphoric highs as far back as the heyday of ancient Egypt. It appears to have been introduced into the United States in the early 1900s by Mexican migrant workers, Latin American mariners, and Caribbean island emigrants. The famed pseudodocumentary film *Reefer Madness* unintentionally reflects the hysteria ensuing from federal government efforts to control the drug's use. However, it was cultivated extensively during World War II due to loss of Asian hemp sources. Its growing use in the postwar years—expanding from underground artists to the 1960s youth subculture—caused a backlash on the part of middle America fueled by opportunistic politicians and law enforcement officials.

Marijuana's notoriety as a popular recreational drug has tended to obscure its alleged medicinal benefits. Marijuana (particularly the species *Cannabis sativa* and *Cannabis indica*) was employed by societies worldwide as far back as antiquity to alleviate physical pain. The herb also served as a conventional treatment for the prevention and relief of migraine headache, a condition known to affect an estimated twenty-three million Americans alone.

Evidence that mainstream physicians once held marijuana in high regard as a medicinal is widely documented in the literature. The herb was listed in the *United States Pharmacopeia* during the years 1860 to 1941. Regarding migraine treatment, the acknowledged father of modern medicine, Sir William Osler, stated, "*Cannabis indica* is probably the most satisfactory remedy."

Despite this demonstrated analgesic value, and vigorous protest by the American Medical Association, marijuana was made illegal in the United States in 1937. The editor of the *Journal of the American Medical Association,* however, maintained support for *Cannabis* in treating menstrual migraine.

Evelyn Leigh notes a revival of interest in the medicinal use of marijuana during the past two decades.

> A study investigating the effects of oral doses of the *Cannabis* compound delta-9-tetrahydrocannabinol in patients with cancer demonstrated a trend toward pain relief with escalating doses. Recent studies have pointed out some possible mechanisms of action for cannabinoids in migraine (interference with pain transmission) in an area of the brain that is considered a likely area for migraine generation.

Prominent health care publications such as *JAMA* and the *British Medical Journal* have also advocated serious examination of the utilization of medical marijuana. Experts have gone public advocating its use as an anticonvulsant, as a muscle relaxant in spastic disorders, as an appetite stimulant in the wasting syndrome of human immunodeficiency virus infection, and in treating glaucoma.

Within the legal arena, U.S. Drug Enforcement Agency administrative law judge Francis Young concluded in 1988, "By any measure of rational analysis marijuana can be safely used within a supervised routine of medical care" (Leigh, Web site, 2000). A number of western states have spearheaded ballot initiatives aimed at the legalization of the herb within a medical context. In short, the general trend toward making medical marijuana more widely available to the public appeared to be gathering momentum at the outset of the twenty-first century.

BIBLIOGRAPHY

Heinerman, John. "Marijuana *(Cannibis sativa).*" In: *Heinerman's Encyclopedia of Healing Herbs & Spices*. Paramus, NJ: Prentice-Hall, 1996, pp. 321-324.

Lee, Martin A., and Shlain, Bruce. *Acid Dreams: The CIA, LSD and the Sixties Rebellion*. New York: Grove, 1985.

Leigh, Evelyn. "Research Reviews: *Cannabis* for Migraine." *Herb World News Online*. Boulder, CO: Herb Research Foundation, 2000. 2p. www.herbs.org/current/cannabis.html.

Robinson, Rowan. *The Great Book of Hemp*. Rochester, VT: Park Street Press, 1996.

Russo, E. "Cannabis for Migraine Treatment: The Once and Future Prescription? An Historical and Scientific Review." *Pain 76* (1998): 3-8.

MELALEUCA

Generally classified as a large bush or small tree, melaleuca *(Melaleuca alternifolia)* is common to the Philippines, Indonesia, and Malaysia, where it is known as the cajeput tree, as well as Australia, where it is called the tea tree. It possesses a dense, narrow crown and a stout, frequently twisted trunk. The papery bark peels off in layers while the thin, leathery leaves are alternate, lanceolate, and pointed at both ends. The finely haired flowers are white and fragrant; they produce a small, grayish-brown fruit distinguished by a groove around the top encompassing a craterlike cup bearing five radial grooves.

In Asia, the melaleuca is used to line streets for shade and ornamental purposes. It is also prized for firewood; in addition, the Malaysians have utilized the bark as a caulk for boats.

An oil imbued with a turpentine-like odor, however, has proven to be the most marketable by-product of the plant. John Heinerman notes that an upscale animal grooming service based in Utah employs melaleuca oil to help heal wounds and sores on clients' pets. The owner has used it on her own wounds to promote fast healing while avoiding infection and swelling. The oil is also used to clean the ears of pets and is a deterrent for fleas and ticks. She considers it "a miracle remedy from Mother Nature" (p. 329).

Indonesian folk healers and scientists have discovered a wide range of applications for melaleuca oil. A small amount added to lukewarm water and bottle-fed to an infant relieves even extreme cases of colic. When rubbed on the forehead and the back of the neck, it terminates the pains associated with migraine headaches. Applications of melaleuca also effectively treat earaches, sore teeth, leg cramps, bedsores, herpes lesions, and outbreaks related to sexually transmitted diseases.

Melaleuca is not without its downside. An article published in the April 21, 1991, issue of *The Orlando Sentinel* (pp. K-1, K-4) termed it "one of the three most environmentally dangerous plants" threatening the entire state. It stated that these "monster" trees were "sucking up water faster than a thirsty construction worker . . . depleting neighboring plants' water supplies, and turning marshes into dry land" (cited in Heinerman, 1996, p. 328). One nature enthusiast, who admitted having been previously opposed to destroying anything occurring naturally within the environment, observed that "they're just taking over and they've got to be stopped."

BIBLIOGRAPHY

Heinerman, John. "Melaleuca *(Melaleuca alternifolia)*." In: *Heinerman's Encyclopedia of Healing Herbs & Spices*. Paramus, NJ: Prentice-Hall, 1996, pp. 328-330.

"Melaleuca—Tea Tree Oil." *HerbMed.* Alternative Medicine Foundation, 1998. 10p. www.herbmed.org/herbs. Includes hyperlinks to articles cited in Medline abstracts.

"Tea Tree Relatives." *The Herb Companion* (October/November 1996): 60.

MILK THISTLE

A hardy annual or biennial herb found in North and South America (particularly the West Coast of the United States and the pampas of Argentina) and southwestern Europe eastward to Afghanistan, milk thistle *(Silybum marianum)* is considered a weed by some due to its invasive character, while others hail its nutritional, medicinal, and decorative qualities.

Marianum is Latin for "of Mary"; legend has it that the white marbling on the milk thistle's wavy, shiny green leaves arose when the Virgin Mary spilled milk on a leaf while nursing the baby Jesus. The leaves are also distinctive for the sharp spines at the lobe tips. The lower leaves, which are stalked, mature to eighteen to thirty inches in length and six to twelve inches in width; those on the flower stalk are progressively smaller and clasp the stem. During the spring of the plant's second year, stiff, erect flower stems reach as high as seven feet. By early summer, two-inch–wide heads of faintly scented red-purple, tubular florets appear on the stem tips above spiny bracts. The flowers act as a magnet for bees, and the subsequent shiny black fruits attract goldfinches and black-headed grosbeaks.

While often used by garden designers, cultivation by hobbyists is prohibited in locales where it is considered an annoying weed. Therefore, it is advisable to contact a nearby agricultural extension office or other agency prior to obtaining milk thistle seeds or plants.

Milk thistle has a long history of culinary use. The peeled, chopped stems are used in salads or soaked (to remove bitterness) and stewed in much the same manner as rhubarb. The leaves (when trimmed of prickles) and roots can be eaten raw or cooked. Maud Grieve, in *A Modern Herbal* (1931), noted that the boiled roots and stalks "surpass the finest cabbage" (cited by Strauch, 2000, p. 14). Furthermore, the flower buds have been prepared like artichokes and the seeds roasted for a coffee substitute.

Based upon the "Doctrine of Signatures," the whitish leaf variegation implied that the plant could increase milk flow; therefore, it became part of the diet of human and animal mothers. A number of problems arose from this practice. Unless the plant tops were simply steeped to make a tea, the spines had to be removed before eating. Because milk thistle accumulates nitrates from the soil, plants grown in nitrate-rich soils could poison livestock feeding on them.

Medicinal uses of milk thistle were even more diversified in antiquity. Ancient herbalists considered the boiled tops of the plant to be an effective spring blood cleanser. They prescribed a drink derived from its seeds as an antidote to snakebite and rabies. In addition, the root was worn so as to "expel melancholy."

In the early 1800s, a noted German physician named Rademacher created a tincture derived from milk thistle seeds for treating liver ailments. The medication, Tinctura Cardui Mariae Rademacher, can be found in present-day European pharmacopeias. The literature concerning milk thistle seed—and its active principle, silymarin—has reached, in John Heinerman's words, "incredible proportions" (p. 331). It remains *the* preferred liver medicine in most of Europe.

Despite the dubious value of some early remedies, scientific studies underscore the medicinal value of the plant. Silymarin has been found to protect the liver by preventing the entry of toxins through cell membranes as well as stimulating the repair of damaged tissue by boosting protein synthesis. Concentrated silymarin preparation in capsule form can effectively treat hepatitis and cirrhosis. Animal experiments indicate that concentrated dosages are an antidote to poisoning by the Death Cap mushroom. Its demonstrated value is such that the herb is actively cultivated in some states, most notably California and Texas. Discount Natural Herbs, a major Internet retailer, ranked milk thistle one of its "Top Ten Selling Herbs" as of mid-2001.

A wide range of studies concerning other possible health benefits of milk thistle are presently in progress or, as yet, inconclusive. Among the issues being considered:

- Whether its addition to bread products increases the consumers' capacity for work
- Whether (and under what conditions) silymarin lowers serum cholesterol
- How silymarin protects the kidneys from injury
- The plant's antioxidant and anticancer properties

BIBLIOGRAPHY

Grieve, Maud, and Leyel, Hilda (Eds.). *A Modern Herbal.* 2 vols. New York: Hafner, 1967, c. 1931.

Heinerman, John. "Milk Thistle *(Silybum marianum).*" In: *Heinerman's Encyclopedia of Healing Herbs & Spices.* Paramus, NJ: Prentice-Hall, 1996, pp. 330-331.

Strauch, Betsy. "An Herb to Know." *The Herb Companion 12*(6) (August/September 2000): 14-15.

MINT FAMILY

The genus *Mentha* includes eighteen species and hundreds of varieties and cultivars. Two mints, however, have long dominated the medicinal field: spearmint *(Mentha spicata)* and peppermint *(Mentha piperita).*

Spearmint, heavily pervaded by scent of carvone, possesses the longest history of medicinal use. The plant was allegedly cultivated in every convent garden in Europe by the nineteenth century. A tea brewed from its leaves was used to treat digestive problems. "Aqua mentha sativa" (known as spearmint water and made by distilling the leaves with three times their weight in water over low heat) was touted by *The Edinburgh New Dispensatory* (1789) as an "incomparable remedy for strengthening a weak stomach and curing vomiting proceeding from cold viscous phlegm."

Peppermint, distinguishable from spearmint by its menthol fragrance, would go on to become the preeminent medical mint in the Western world. It was recommended as early as 1704 in Ray's *Historia Plantarum* for the treatment of "stomach weakness" and diarrhea. Accorded official status in the *London Pharmacopoeia* (1721), the plant was being systematically cultivated in England by the mid-eighteenth century.

Peppermint was widely used by early European settlements in the New World. The Shakers employed distilled peppermint water as a digestive cordial at the Sabbathday Lake Shaker Community (Maine) in the late 1700s. Samuel Stearn's *American Herbal* (1801) noted that the plant "restores the functions of the stomach, promotes digestion, stops vomiting, cures the hiccups, flatulent colic, hysterical depressions, and other like complaints." The *U.S. Dispensatory,* published in 1848, referred to peppermint as "a very grateful aromatic stimulant" and cited its value in allaying nausea, relieving spasmodic pains of the stomach and bowels, expelling gas, and covering the taste of other medicines. Furthermore, the fresh herb was argued to be an effective means of combating cholera in youth.

The increasing popularity of peppermint had severely undercut spearmint's use by the outset of the twentieth century. The eighteenth edition of King's *American Dispensatory* (1905) observed that the latter was "somewhat inferior to peppermint" as a carminative, antispasmodic, and stimulant. In 1812, peppermint was cultivated for oil production in Ashfield, Massachusetts; by midcentury, the United States was a leading producer of mint oil with major growing centers in Michigan, New York, and Ohio. In the late 1990s, nearly 100,000 acres in Idaho, Indiana, Michigan, Oregon, Washington, and Wisconsin remained devoted to the industry.

The German Commission E, responsible for regulating herb usage in that nation, has verified the effectiveness of fresh or dried peppermint leaf for treating spastic disorders of the gastrointestinal tract, gallbladder, and bile duct as well as for relieving gas in the digestive system. Scientific research has also indicated that peppermint oil inhibits the growth of bacteria and relieves spasms of the digestive tract. Physicians also prescribe it for the treatment of irritable bowel syndrome (IBS)—a chronic condition characterized by abdominal pain, bloating, and constipation, or even diarrhea—and inflammation of the oral mucosa. Spearmint, on the other hand, is now limited primarily to herbal teas for its flavoring properties.

While continuing to be widely used in Europe in over-the-counter preparations, peppermint and peppermint oil were denied OTC drug status in the United States by a Food and Drug Administration review panel in 1990 in the absence of new data to support their efficacy and safety. Where still employed, potential consumers are advised not to utilize peppermint-based products for inflammation of the bile duct or gallbladder or for severe liver damage. Peppermint tea poses problems for those afflicted with gastroesophageal reflux disorder; by relaxing the sphincter at the upper end of the stomach, it causes stomach acid to move into, and further irritate, the esophagus. The menthol vapors in both peppermint tea and oil are also reputed to cause choking in infants and small children.

Nevertheless, mint remains one of the top-selling botanicals. This is underscored by its inclusion in Deborah C. Harding's *The Green Guide to Herb Gardening: Featuring the 10 Most Popular Herbs* (2000).

BIBLIOGRAPHY

Foster, Steven. "Medical Mints." *The Herb Companion* (August/September 1997): 57-59.

Foster, Steven. *Peppermint (Mentha piperita).* Botanical Series, No. 306, Second edition. Austin, TX: American Botanical Council, 1996.

Gardner, Jo Ann. "Mint." In: *Living with Herbs.* Woodstock, VT: Countryman Press, 1997, pp. 199-205.

Harding, Deborah C. *The Green Guide to Herb Gardening: Featuring the 10 Most Popular Herbs.* St. Paul, MN: Llewellyn, 2000.

Landing, J.E. *American Essence: A History of the Peppermint and Spearmint Industry in the United States.* Kalamazoo, MI: Kalamazoo Public Museum, 1969.

Rees, W.D.W., Evans, B.K., and Rhodes, J. "Treating Irritable Bowel Syndrome with Peppermint Oil." *British Medical Journal* 2(6194) (1979): 835-836.

Somerville, K.W., Richmond, C.R., and Bell, G.D. "Delayed Release of Peppermint Oil Capsules (Colpermin) for the Spastic Colon Syndrome: A Pharmacokinetic Study," *British Journal of Clinical Pharmacology 18* (1984): 638-640.

MOTHERWORT

Names reveal much about this herb. The generic name, *Leonurus,* refers to the supposed resemblance of its shaggy, palmately lobed leaves to lions' tails ("Leon" being Greek for "lion," *-urus,* Latin for "tail"). Its specific name, *cardiaca* (Latin for "relating to the heart"), takes into account the plant's longtime role as a heart medicine. And its popular name reflects an equally long history of use in easing childbirth.

Motherwort's stout, squared stalks arise from a basal rosette of dull green leaves, which are reminiscent of maple leaves. The number of lobes possessed by the leaves vary according to placement; those lower on the stem have five, those located in the middle have three, and those near the top—small and narrow in appearance—have one or none at all. The leaves are also distinguished by a strong, pungent odor and a network of veins most readily discerned on the pale undersurface.

During summer months, tiny, two-lipped flowers (pink, white, or purple in color) bloom in clusters around the upper stems at the leaf nodes. Each flower goes on to produce four three-sided black seeds.

Native to Europe, motherwort was brought to the New World as a medicinal plant by the early colonists. However, it soon escaped cultivation and can now be found growing wild across lower Canada and most of the United States.

Betsy Strauch notes that herbalists worldwide throughout history have considered the plant "virtually a panacea" (p. 12). Typically administered with a sweetening agent to help mask the bitter taste, it has been used as a sedative or tonic, rabies cure, treatment for asthma, diuretic, and stimulant to contract the uterus following childbirth or to bring on a delayed menstrual period. Scientists have ascribed its effectiveness as a uterine tonic to its alkaloid content; likewise, its glycoside content aids the heart, and its tannin content, the digestive tract.

Traditional Chinese Medicine has displayed a similar fixation with a close relative, *Leonurus sibiricus*. The tops are used to lower blood pressure and stimulate the uterus and kidneys, while the seeds are part of the treatment plan for certain eye disorders.

Motherwort does have its downside. The cucumber mosaic virus, a condition affecting cucumber, summer squash, and tomato plants (distorting leaves and rendering the fruit bitter), can winter over in motherwort. During the next growing season, aphids or cucumber beetles carry the virus from the motherwort to susceptible vegetables. In addition, some people acquire a skin rash after handling it.

BIBLIOGRAPHY

Heinerman, John. "Motherwort *(Leonurus cardiaca).*" In: *Heinerman's Encyclopedia of Healing Herbs & Spices*. Paramus, NJ: Prentice-Hall, 1996, p. 339.

Strauch, Betsy. "An Herb to Know." *The Herb Companion 13*(1) (October/November 2000): 12.

N

THE NEW WORLD TAPESTRY AND HERBS

The New World Tapestry is a 264-foot-long tapestry project that highlights the colonizing of North America and the plants that came with the settlers. It is displayed in a revitalized woolen mill in Uffculme, Devon, England. An herbalist's delight, it is composed of twenty-three panels, each eleven feet by four feet in size. It utilizes a tent stitch, or Oblique Gobelin stitch, sixteen stitches to the inch in wool generously donated by Emu Wool, among others. The overall tapestry design is colorful strip cartoons portraying colonial life between 1583 and 1642, including emigrant families, their companions, and the American Indians. The tapestry represents the largest display of needlework in existence, according to the man who conceived its design.

The project was created in 1975 when Plymouth, England, artist and advertising executive Tom Mor wanted to honor the city from whose harbor the Mayflower and other famous expeditionary vessels had embarked to North America. Once he finished his tapestry design, however, more practical concerns intruded on his idea: no public building in Plymouth was large enough to display the finished work. In 1995, after the first ten panels were ready, Mor arranged for their permanent home in Uffculme's vast, run-down, and abandoned Coldharbour Mill, located seventy miles from Plymouth but within two miles of a superhighway linking England's West Country to London. Located in Devon's verdant Culm Valley, where the once-traditional enterprises of farming and of light industry have declined, the tapestry provides a "stitch-in-time" tourist attraction, especially to American history buffs, needlework enthusiasts, and amateur botanists interested in historic plants and herbs transplanted from England to the New World.

The panels illustrate the joys and tribulations of the early North American settlers along with the colonies of Bermuda and of Guiana. Swashbuckling adventurers, such as Sir Walter Raleigh and Sir Francis Drake, along with such romantic figures as Miles Standish, Captain John Smith, and Pocahontas, dominate sketches of American colonial life; coats of arms proclaim personal status in the home country; and depictions of the herbs they used for sustenance and medicine provide a historic link between the two countries. The herb designs, taken from old herbals, such as those written by Gerard, Culpeper and Parkinson, represent the period with which many of the characters displayed in the tapestry would be familiar. When finished, the upper and lower borders formed an imaginative display of plant history while the center panel will be a plan of the Oxford Botanic Garden, which was the first to be founded in England (1621).

The New World Tapestry was more than a decade in the making and contains nearly thirty-nine million stitches in primary color yarns. It reflects the collective efforts of accomplished volunteers who worked originally in nine stitching centers throughout Devon and Dorset, but now create their designs through five main centers in Plymouth, Exeter, Bideford, Lyme Regis, and Tiverton, as well as contributions of stitches from patrons worldwide. Each stitch costs more than one English pound; a certificate is given to each participant to verify their respective contributions. Certain panels bear the gold needlework of members of the Royal Family, including H.M. Queen Elizabeth and the Duke of Edinburgh, who each made their own stitch in 1988 during the Armada celebrations at HMS Drake at Devenport. Honorary patrons include the governors of some of the New England states and the U.S. Ambassador to the United Kingdom.

To accommodate its new fame, Coldharbour Mill has been refurbished with a working waterwheel, restored engine houses, a weaver's and a dyer's workshop, a tearoom, and a pastoral picnic area along the mill pond.

BIBLIOGRAPHY

McGrath, Jacqueline. "A Herbalist's Delight: The New World Tapestry." *Herb Quarterly 86* (Summer 2000): 5.

NUTMEG

A product of the seed of the nutmeg tree—an evergreen capable of reaching more than sixty feet in height with spreading branches and dense foliage—nutmeg *(Myristica frangrans),* while best known as a culinary flavoring, is also used for medicinal purposes. The tree's fruit is two to four inches in length and possesses a fleshy character reminiscent of an apricot. It splits in half when ripe, exposing a seed inside a dark reddish-brown, brittle shell. The netlike aril surrounding the shell is mace, which turns a yellowish- or orange-brown upon drying.

Native to the Moluccas Islands of Indonesia, nutmeg is presently grown throughout the tropics. Commercial cultivation, however, is centered in Indonesia and Granada, an island within the West Indies.

Nutmeg is employed in the flavoring of meat sauces, tomato ketchup, and vegetable juices as well as cola drinks. The kernels deemed undesirable for culinary applications due to damage or insect attack are used for distillation. It is also an ingredient in perfumery and toiletry products, most notably spice aftershaves for men.

Nutmeg has been popular for centuries in treating stomach ailments such as gas, indigestion, nausea, and vomiting as well as kidney problems. One form of medication includes a pasty blend of slippery elm bark, nutmeg, mace, and cold water combined with heated half-and-half.

In recent decades, the spice has also been utilized by physicians and psychiatrists as an alternative to psychedelic drugs for patients in need of a nonaddictive, mind-altering substance that assists them in coping with reality. Professional journals such as the *Journal of Neuropsychiatry* (March-April 1961) and the *New England Journal of Medicine* (August 22, 1963) have noted that one tablespoon of powdered nutmeg is sufficient to impart a "dream-like, floating and slightly euphoric" sensation (cited in Heinerman, 1996, p. 351). Furthermore, the user can revert back to reality whenever he or she chooses to do so. A psychiatrist at the University of Maryland School of Medicine—in an angry mood due to a particularly irritating problem—noted, after taking the spice, that "the anger was dispelled and I felt at peace with the world. I wandered out to a leisurely lunch with some friends and felt quite unconcerned about my work. This is unusual for me" (Heinerman, 1996, p. 382).

The use of nutmeg does result in a number of physical discomforts, albeit not life-threatening ones. These include aching muscles and bones, draining sinuses, eye pain, and limited diarrhea. The general consensus from medical practitioners, however, is that the therapeutic benefits more than compensate for these inconveniences.

BIBLIOGRAPHY

Heinerman, John. "Nutmeg *(Myristica fragrans).*" In: *Heinerman's Encyclopedia of Healing Herbs & Spices*. Paramus, NJ: Prentice-Hall, 1996, pp. 350-353.
"Nutmeg Oil." In: Charalambous, George (Ed.), *Spices, Herbs and Edible Fungi.* Amsterdam: Elsevier, 1994, pp. 36-37, 150-178.

O

ONION

The onion *(Allium cepa)* belongs to the lily family. Similar to close relations such as garlics, leeks, chives, scallions, shallots, and asparagus, it contains an abundance of the trace element compound sulfur, which is responsible for its distinctive flavor.

The popularity of the vegetable can be documented as far back as antiquity. After being liberated from their Egyptian captivity by Moses, the Hebrews complained on more than one occasion: "we remember the fish, which we did eat in Egypt freely; the cucumbers, and the melons, and the leeks, and the onions . . . " (Numbers 11:5, *Old Testament*). The Roman historian Pliny wrote that the Egyptians were known to invoke the onion when taking oaths as is presently done with the Bible during legal proceedings.

The vegetable would later play a role in the naming of the third largest city in the United States. In 1624, French explorer Pere Jacques Marquette was shown thousands of acres on the southern shore of Lake Michigan overgrown with wild onions. His Indian guides uttered a word which denoted the offensive odor which permeated the region: "Chicago!"

The presence of hundreds of varieties of onions in the marketplace today underscores the vegetable's widespread appeal as a culinary delicacy. The healthful and medicinal benefits of onions, however, are not nearly as well known by the general public.

Belle Boone Beard, a sociologist affiliated with the National Institute on Aging, noted that nearly all of the 8,500 centenarians whom she'd surveyed ate onions just about every day. Given that the vegetable is a strong antioxidant and can control free radicals tied to the aging process, it would appear that onions help impede the inroads of old age.

A sixteenth-century French surgeon, Ambroise Pare, adapted a popular folk remedy dating from the Middle Ages to treat a kitchen boy's severe burns. He relates that "I the next day found those places of his body whereto the Onions lay, to be free from blisters, but the other parts which they had not touched, to be all blistered." Pare continued to enjoy the same success when employing this treatment on soldiers and other patients suffering from all degrees of burns.

A syndicated Ann Landers column published a beekeeper's regimen for treating bee stings: "I applied the onion and lo and behold, as if by magic, the pain stopped within minutes and the swelling went down."

British herbalist Mary Thorne Quelch believed that sliced onion absorbed the disease germs causing smallpox, diphtheria, measles, mumps, chicken pox, scarlet fever, tuberculosis, and other epidemics. Likewise, General Ulysses S. Grant was of the opinion that the vegetable could prevent dysentery and other illnesses associated with warm climates.

Based upon the studies of guinea pigs outlined in the May-June 1984 issue of *Agents Actions* (Volume 14, pages 626-629) (cited in Heinerman, p. 361), an alcoholic extract of onions seems to be effective in treating asthma or bronchitis attacks. John Heinerman claims all asthma and bronchitis cases taking his onion-brandy mixture no longer experienced breathing-related problems.

BIBLIOGRAPHY

Ann Landers. *Salt Lake Tribune,* December 15, 1986, A-14.

Farrell, Kenneth T. *Spices, Condiments, and Seasonings.* Westport, CT: AVI, 1985.

Garland, Sarah. *The Complete Book of Herbs & Spices.* Pleasantville, NY: Reader's Digest Association, 1993.

Heinerman, John. "Onion *(Allium cepa).*" In: *Heinerman's Encyclopedia of Healing Herbs & Spices.* Paramus, NJ: Prentice-Hall, 1996, pp. 357-362.

Sigerist, Henry E. "Ambroise Pare's Onion Treatment of Burns." *Bulletin of the History of Medicine 15* (February 1944): 143-149.

OPIUM POPPY

An annual easily propagated from seed in most climates, the opium poppy *(Papaver somniferum)* prefers rich, moist soil and ample sunshine. Typically reaching four feet in height, it produces large, pale mauve or white flowers from midsummer. Originally grown in the Middle East during antiquity, its cultivation had spread as far as China before A.D. 1,000.

The drug for which the plant is best known is obtained by making an incision in the unripe seed capsule and collecting the thick juice that oozes out once it has been dried by the sun. In warm locales, harvested opium is a potent substance. Its use for recreational purposes is illegal in most countries. Opium was used as a medicine in the ancient civilizations of Egypt, Greece, and Rome. Medicinal applications are still permitted today in certain controlled situations. This is also true of two highly valued derivatives, codeine and morphine.

Ripe seeds of the plant do not possess narcotic properties. They are widely used to flavor breads, cakes, curries, and noodles. A by-product, poppy seed oil, is a fixture in cooking.

BIBLIOGRAPHY

Garland, Sarah. *The Complete Book of Herbs & Spices*. Pleasantville, NY: Reader's Digest Association, 1993.

Griggs, Barbara. *Green Pharmacy: The History and Evolution of Western Herbal Medicine*. Rochester, VT: Healing Arts, 1997.

P

PARSLEY

A biennial or short-lived perennial that matures to two feet or more in height, parsley *(Petroselinum crispum)* appears in numerous varieties. It possesses a thin, white, spindle-shaped root that produces an erect, grooved, nonhairy, multibranched stem. During summer months, it produces white or green-yellow flowers in compound umbels. Useful parts include the seeds, leaves, and (chopped) stems.

Parsley is widely cultivated for its foliage, particularly in Belgium, France, Germany, Hungary, and California. Although poisonous to most birds, it has many beneficial aspects for animals; for example, curing foot rot in sheep and goats.

While best known as a seasoning and visual accent in food preparation, parsley has also possessed many medicinal uses throughout history. Theophrastus called it the coronary plant, believing the plant to be good for treating coronary conditions. Dioscorides claimed that consuming it would provoke urine. Ancient writers claimed parsley increased the blood, soothed the heart and stomach, and warded off intoxication; for this reason, Greek and Roman heroes wore it on their heads at ceremonial banquets. During the Middle Ages, the Anglo-Saxons used wild parsleys to mend skulls that had been fractured in combat. An age-old Romanian gypsy remedy consisted of applying crushed sprigs of parsley leaves to skin bruises in order to clear up black-and-blue marks. Natives of Mexico's Yucatan Peninsula have long used a tea brewed from chopped parsley to treat kidney inflammation, inability or pain in urination, kidney stones, and edema.

Parsley tea also has been employed with some degree of success as a mild aphrodisiac for couples plagued by sexual frigidity. Furthermore, the plant has been fed to sheep in Spain to bring them into heat at any time of year.

Parsley contains high concentrations of vitamins A and C, both of which are scientifically proven to be significant nutrients in fighting cancer. Furthermore, the plant outranks most vegetables in the amount of histidine, an amino acid that strongly inhibits tumor development. Its inclusion in Deborah C. Harding's *The Green Guide to Herb Gardening: Featuring the 10 Most Popular Herbs* (2000) reflects its continued popularity in the present day.

BIBLIOGRAPHY

Charalambous, George (Ed.). *Spices, Herbs and Edible Fungi.* Amsterdam: Elsevier, 1994.

Farrell, Kenneth T. "Parsley." In: *Spices, Condiments, and Seasonings.* Westport, CT: AVI, 1985, pp. 74-77.

Harding, Deborah C. *The Green Guide to Herb Gardening: Featuring the 10 Most Popular Herbs.* St. Paul, MN: Llewellyn, 2000.

Heinerman, John. "Parsley *(Petroselinum crispum).*" In: *Heinerman's Encyclopedia of Healing Herbs and Spices.* Paramus, NJ: Prentice-Hall, 1996, pp. 363-366.

Hollis, Sarah. "Parsley." In: *The Country Diary Herbal.* New York: Henry Holt, 1990, pp. 80-82.

PAU D'ARCO

Pau d'arco *(Tabebuia species)* is a large, deciduous tree native to the South American rain forests. It reaches approximately 100 feet in height and often possesses a trunk circumference of more than four feet. It produces groupings of five elliptic leaves that are either lanceolate or ovate shaped. Residing on stems that are dark green on top and yellow-green below, the leaves are placed in whorls comparable to fingers on a human hand. Varying in color according to species, the flowers range from blue and yellow to magenta and purple. They are either bell- or funnel-shaped with yellow throats.

Used for centuries in South American countries, pau d'arco has been a fixture within the North American health food industry for a couple of decades as a folk treatment for various cancers. It is available in tea, tincture, or capsule forms.

Research studies have shown that the herb contains a natural antibacterial agent, cleanses the blood, and kills viruses. These features serve to build the immune system and increase resistance to disease. The conditions it can effectively treat include bronchitis, osteomyelitis, ringworm, colitis, cystitis, prostatitis, lupus, diabetes, eczema, liver disease, snail fever, and all infections. Michael Tierra, a practicing herbalist based in Santa Cruz, California, states, "Pau d'arco's chief purpose seems to be in the elimination of pain caused by almost any disease you can think of" (Heinerman, 1996, p. 367).

An article published in the *Journal of Herbs, Spices & Medicinal Plants* (Volume 2, Number 4; 1994), found that most commercial products utilizing the herb are not very potent or effective (cited in

Heinerman, 1996, p. 368). Analysis of a dozen such products determined that ten lacked sufficient amounts of lapachol—an important naphthoquinone—to produce the desired medical result. Customers of health food retailers and direct marketing companies have been advised in the mass media to check the ingredients of a particular product prior to purchase. Nevertheless, pau d'arco remains widely used today, as evidenced by its inclusion in Discount Natural Herbs' "Top Ten Selling Herbs."

BIBLIOGRAPHY

Heinerman, John. "Pau D'Arco *(Tabebuia species)*." In: *Heinerman's Encyclopedia of Healing Herbs & Spices*. Paramus, NJ: Prentice-Hall, 1996, pp. 367-368.

PEACH OIL

Researchers worldwide have examined natural chemicals for many years in an effort to find substitutions for synthetic compounds feared for their dangerous side effects. Some of these scientists—most notably, teams based at the U.S. Agricultural Research Service and in South Africa and Israel—have determined that peach oil (the substance providing the fruit with its characteristic fragrance) kills fungi and other pests residing in the soil. This byproduct, benzaldehyde, not only destroys *Fusarium oxysporum, Rhizoctonia solani, Pythium aphanidermatum,* and *Sclerotinia minor,* but apparently favors other organisms in the soil that muscle out unwanted pathogens.

Benzaldehyde is already being manufactured synthetically, and is in use commercially, much like similar oils distilled from lemon and peppermint. The essence of peach is being touted as an alternative to methyl bromide, a widely employed pesticide that is toxic to many life forms and destructive to the earth's protective ozone layer.

Implementation of such natural substances, however, is an uphill battle. Charles L. Wilson, a plant pathologist at Agricultural Research Service's Appalachian Fruit Research Station in Kearneysville, West Virginia, states, "There is such a broad range of these things that are in nature already, and there has not been that much effort in trying to fish them out. The synthetic compounds were so powerful that we stopped looking for natural pesticides" (Johnston, 2000, p. 24). Thomas Duafala, a research chief at the Hollister, California-based Trical, a company that applies pesticides on farms, estimates that it would take years to get regulatory approval for any new pesticide.

Furthermore, the use of new substances typically triggers concerns not anticipated prior to implementation. Tom Kurt, a physician on the American Botanical Council Advisory Board, notes that individuals coming into contact with benzaldehyde during commercial clean-ups have had allergic reactions.

BIBLIOGRAPHY

Cushman, John H. Sr. "Peach Oil May Work as Pesticide." *The New York Times,* March 14, 1999.

Johnston, Barbara A. "Peach Oil as Pesticide?" *HerbalGram 48* (2000): 24. www.herbalgram.com.

PSYLLIUM

A short-stemmed (sometimes stemless) annual, psyllium *(Plantago ovata)* produces leaves—ranging from three to ten inches in length and up to one-half inch in width—in a rosette or alternate arrangement which clasp the stem in straplike fashion. Its minute, white flowers are four-parted, in erect, ovoid, or cylindrical spikes. The top half of the fruit separates when ripe, releasing smooth, ovate seeds colored either pinkish-gray-brown or pinkish-white with brown streaks. Encased in a thin, white, translucent husk, the seed expands greatly when soaked in water.

Psyllium was introduced into Europe during the Middle Ages by Arab pharmacists. It was extremely popular with medieval doctors, who focused on the evacuation of excess humors via purges, cathartics, and laxatives.

Research has demonstrated that psyllium outperforms other known laxatives such as milk of magnesia, mineral oil, methylcellulose, phenolphthalein, or cascara sagrada. Many popular bulk laxatives (e.g., Effersyllium, Metamucil, Syllamalt) are composed of ground psyllium husks or seeds laced with sugar to improve the taste.

Italian clinical experiments have shown a significant decrease in serum cholesterol as well as reduced food intake on the part of obese subjects who ingested psyllium seeds. Furthermore, diabetic patients experienced a lowering of blood sugar levels. In southern California, physicians found that the herb is effective in relieving irritable bowel syndrome. All of these conditions can now be treated by a product called Fiber Cleanse.

BIBLIOGRAPHY

Griggs, Barbara. *Green Pharmacy: The History and Evolution of Western Herbal Medicine*. Rochester, VT: Healing Arts, 1997.

Heinerman, John. "Psyllium *(Plantago ovata)*." In: *Heinerman's Encyclopedia of Healing Herbs & Spices*. Paramus, NJ: Prentice-Hall, 1996, pp. 380-381.

PURPLE LOOSESTRIFE

Native to Europe, purple loosestrife *(Lythrum salicaria)* now thrives in wetland areas throughout New England and the East Coast. Reaching two to four feet in height, it grows brilliant purple flower spikes from July through September. Sometimes named spike loosestrife, it has been widely criticized by ecologists because it is aggressive and tends to take over in a very short period of time. However, the plant does possess many beneficial features.

Purple loosestrife has a rather unique blend of astringent and demulcent properties. It is an anti-inflammatory, and a rich source of both tannins and antioxidant compounds called proanthocyanidins.

The plant has a long history of use in Europe as a gargle for sore throats (particularly those related to fevers), a lymphatic cleanser, a remedy for bacterial and amoebic dysentery, and a wash for wounds.

Purple loosestrife may be best known among health care specialists today as a treatment for various gastrointestinal problems, most notably Crohn's disease, irritable bowel syndrome (IBS), and leaky gut syndrome. The proanthocyanidins play a key role in strengthening arterial walls and the gastrointestinal lining. The tannins act as an astringent to counter diarrhea, while irritated mucous membranes typical of IBS and Crohn's disease are soothed by the plant's demulcent features. In addition, purple loosestrife is sometimes used as an eyewash for conjunctivitis and macular degeneration, and is known to lower blood glucose levels, a characteristic that may render it a viable antidote to Type II diabetes.

BIBLIOGRAPHY

Langer, Robin E. "Herb-of-the-Month: Purple Loosestrife *(Lythrum salicaria)*—August 1998." *Snowbound Herbals* (August 1998): 2p. www.sbherbals.com.

R

RED CLOVER

Red clover *(Trifolium pratense)* is prevalent in fields and waste areas; its very availability renders it virtually unnoticeable. It possesses a mildly sweet taste and has long been employed as cattle fodder. It is also planted to replenish the soil's nitrogen content when rotating crops.

Red clover generally blooms from July through September. The blossom is the portion of the plant used in medicine. Red clover's active constituents include coumarins, cyanogenic glycosides, flavonoids, phenolic glycosides, salicylates, and mineral acids. It has been used as an expectorant and antispasmodic for coughs, most notably those related to asthma and bronchitis; to purify the blood; to strengthen liver functions; to treat cancer (e.g., the Hoxey Formula); and to relieve chronic skin conditions such as eczema, acne, and psoriasis.

The plant's benefits can be realized through application as an ointment (e.g., lymphatic swellings) or a tincture for skin problems. It also can be ingested as a tea or syrup. The tea is sometimes employed in a hot compress to relieve arthritic pains and gout.

The salicylates found in red clover can potentially affect blood coagulation; while it is known to keep the blood from becoming too sticky, no reports of thinning the blood too much have yet been recorded. However, individuals on an anticoagulant regimen involving aspirin or Coumadin (warfarin) are advised to carefully monitor coagulation times under the close supervision of a qualified health care provider.

BIBLIOGRAPHY

Heinerman, John. "Red Clover *(Trifolium pratense)*." In: *Heinerman's Encyclopedia of Healing Herbs & Spices*. Paramus, NJ: Prentice-Hall, 1996, pp. 390-393.

Langer, Robin E. "Herb-of-the-Month: Red Clover *(Trifolium pratense)*—March 1998." *Snowbound Herbals* (March 1998): 2p. www.sbherbals.com.

"Red Clover and Biochanin A." *The Herb Companion* (October/November 1996): 60.

RHUBARB

Rhubarb root has been well known as a medicinal for over 2000 years. *Rheum palmatum,* which originated in northern China and Tibet, arrived in Europe by means of trade through Asia Minor. This movement accounts for its popular name, Turkey rhubarb. This plant reaches considerably greater size (up to eleven feet with a root weighing as much as thirty-five pounds) and has a stronger purgative action than common garden rhubarb (*Rheum rhabarbarum* or English rhubarb), whereas the latter possesses a stronger astringent action. The use of rhubarb stalks for fruit is a comparatively recent phenomenon; the first such documented use in England was in 1810.

For medicinal purposes, the most effective roots are harvested in October at a minimum age of six years. The process involves slicing the roots and drying them first in the open air, and later within heated ovens. The active constituents—anthraquinone glycosides, tannins, calcium oxalate, resins, and minerals—contribute to the bitter taste and cold, drying energy of the root.

In small dosages, the herb assists in improving digestion. Its astringency—derived largely from the tannin content—helps in relieving diarrhea. Increased amounts stimulate the intestinal tract, thereby fighting constipation. Turkey rhubarb is a key ingredient in Renee Caisse's Essiac, which helps the body remove built-up waste products and improves digestion. The popular formula improves symptoms from breast cancer and is also employed for other diseases that are neutralized by healthier detoxifying systems in the body.

Traditional Chinese Medicine has utilized rhubarb to clear "heat" from the liver, stomach, and blood. "Heat" in the stomach is believed to cause acid reflux, ulcers, mouth sores, gum disease, and other conditions, whereas "heat" in the blood is associated with dysentery or high fever symptoms.

BIBLIOGRAPHY

Grieve, M., and Leyel, Hilda (Eds.). *A Modern Herbal, Volume II*. New York: Dover, 1981, pp. 675-680.

Griggs, Barbara. *Green Pharmacy: The History and Evolution of Western Herbal Medicine*. Rochester, VT: Healing Arts, 1997.

Langer, Robin E. "Herb-of-the-Month: Rhubarbs (*Rheum palmatum* and *R. rhabarbarum*)—March 2000." *Snowbound Herbals* (September 6, 2000): 2p. www.sbherbals.com.

ROSEMARY

An evergreen shrub, rosemary *(Rosemarinus officinalis)* originated in the Mediterranean region (where it still grows wild) and is cultivated in temperate climates throughout the world. In ideal situations, it can reach more than six feet in height with a life span of twenty years. Like its close relation lavender, it possesses a twisted and woody stem. The leathery thick leaves are lustrous, dark green above and silvery underneath with a prominent vein in the middle. Pale blue flowers bloom in early summer (other varieties include white and bright-blue flowers) and strongly attract bees.

The importance of rosemary throughout history is reflected in the rich vein of folklore and traditions surrounding its use. Dioscorides wrote that if the herb were boiled in water and drunk prior to exercising it would cure yellow jaundice. According to legend, the Virgin Mary washed her sky-blue cloak and spread it over a rosemary bush to dry; from that point onward the flowers were blue. In ancient Rome, rosemary was a symbol of fidelity as well as being used to flavor wine and treat insect bites. It was employed to perfume women's baths in Turkey, Greece, and France. In the Middle Ages, it found favor as incense for funeral ceremonies. Another legend held that rosemary placed in clothes closets protected apparel from moths and other vermin. French hospitals would burn it to sanitize the air and prevent infections. The oil of rosemary was employed by apothecary dispensers in order to prevent baldness.

As a flavoring agent, rosemary is considered a delicious, powerful herb, capable of dominating most culinary preparations. It is popular with marinades, stews, potatoes, beef, lamb, poultry, strong game dishes, and puddings and custards. It is also used as a garnish for salads and to flavor sugar. Because it is a natural antioxidant, it can serve as a substitute for BHA and BHT to preserve cereal, luncheon meat, pizza, and other foods. Its present-day popularity is such that it was included in the "Top 10 Flavors for 2000" section of McCormick's *Flavor Trends 2000*.

Among its varied medicinal applications, rosemary is employed to sterilize water contaminated by bacteria. *HerbalGram* editor Rob McCaleb believes that boiling potentially unhealthy water with a bit of the herb will protect against cramps, diarrhea, and fever; this practice is recommended when visiting third world nations.

Brewed as a tea, rosemary is an effective mouthwash for bad breath. An oil extract of the herb functions as a liniment for sore muscles and sprains.

French herbalist Maurice Messegue has referred to rosemary as "the miracle herb that restores youth." A tincture combining rosemary, lavender, and brandy has been used as a digestive aid that increases the flow of bile from the liver into the intestines; this elixir also fights rheumatism, gout, kidney stones, and urination difficulties. Crippled by rheumatism and gout in her seventies, Queen Elizabeth of fourteenth-century Hungary tried a similar formula. She regained a significant measure of her youth, and received a marriage proposal from the King of Poland!

The herb has come to be increasingly used as a skin-softening agent. An extract of rosemary oil first appeared in a product known as Coat-So-Soft, which was marketed by the Santa Barbara–based Rio Vista for soothing the coat of horses. Horse owners using the botanical spray began to notice that their own hands felt softer and smoother; many even started applying it to other body parts. John Heinerman claims a number of Avon distributors have told him confidentially that they switched from their own company's Skin-So-Soft to Coat-So-Soft because they liked it better.

Another of Rio Vista's rosemary-based products, Tail & Mane Detangler/Conditioner, is also touted by horse owners as being more effective than ordinary shampoos and conditioners on their own hair. They cite its effectiveness in repairing damaged hair as well as preventing follicles from becoming brittle, making combing easier, and imparting a more luxuriant shine.

BIBLIOGRAPHY

Charalambous, George (Ed.). *Spices, Herbs and Edible Fungi.* Amsterdam: Elsevier, 1994.

Farrell, Kenneth T. "Rosemary." In: *Spices, Condiments, and Seasonings.* Westport, CT: AVI, 1985, pp. 190-194.

Garland, Sarah. *The Complete Book of Herbs & Spices.* Pleasantville, NY: Reader's Digest Association, 1993.

Heinerman, John. "Rosemary *(Rosemarinus officinalis).*" In: *Heinerman's Encyclopedia of Healing Herbs & Spices.* Paramus, NJ: Prentice-Hall, 1996, pp. 403-406.

S

SAFFLOWER

An annual indigenous to the Mediterranean area, Safflower *(Carthamus tinctorius)* features a glabrous, branching stem and alternate, sessile leaves marked with small, spiny teeth and possessing either an oblong or ovate-lanceolate shape. The orange-yellow flowers are often mistaken for saffron, hence the nickname "false saffron." Reaching a height of one to three feet, the plant is widely cultivated in Europe and America.

A naturopathic doctor made the following observation regarding safflower's medicinal value:

> In my private practice I sometimes see cases of hysteria and seizures in different stages of progress. The single most useful thing I have ever found in controlling these conditions is safflower. I prescribe it in the capsule form (4 daily) as well as in the oil (1-1/2 tbsps. every day). (Heinerman, 1996, p. 407)

BIBLIOGRAPHY

Garland, Sarah. *The Complete Book of Herbs & Spices*. Pleasantville, NY: Reader's Digest Association, 1993.

Heinerman, John. "Safflower." In: *Heinerman's Encyclopedia of Healing Herbs & Spices*. Paramus, NJ: Prentice-Hall, 1996, pp. 407-408.

SAFFRON

Saffron *(Crocus sativus)* has the distinction of being the most expensive spice on the planet, bar none. A cultivated bulb doesn't begin flowering until a couple of years have past; after several years of use, the plants—in order to remain useful—must be dug up, cleaned, and replanted. And, of greatest significance, it takes some 150,000 flowers to produce one kilogram of the desired product. As a result, the saffron purchased today in food outlets consists of the dried stigmas of the saffron crocus flower, a plant grown in the Near East and Mediterranean countries (Valencia is believed by experts to have the highest grade). Due to its high price, turmeric or a suitable U.S.-certified dye or food color is substituted in the making of baked goods, confections, paellas, chicken potpies, and other American food products.

A Theban medical papyrus noted the use of saffron in 1352 B.C. It was used, along with cassia and cinnamon, for annointing the Egyptian pharaohs. The poets of classical antiquity were moved to praise its desirable qualities in numerous instances. The spice appears to have been introduced to northern Europe by the Romans; the Muslims carried it to Spain in the eighth century; and the crusaders returning from the Holy Lands reintroduced it to Great Britain. It was one of the most common ingredients recorded in recipes of medieval Europe, most notably fish and meat dishes, puddings, pies, and wine. England's Henry VIII forbade the Irish to dye their linen with saffron in the sixteenth century because the latter believed it had a sanitizing quality and, as a result, wouldn't wash their linen as frequently as they should.

Saffron has a bright orange color and an intensely sweet and pungent odor. As a flavoring, it is best prepared by placing the dried filaments in a little warm water, milk, or cooking liquor; after allowing it a few minutes to color, scent, and flavor the liquid, add the liquid to the dish (straining is optional). The filaments can also be baked to a crisp and then crumbled into food. It is prized as the traditional flavoring of European dishes such as Cornish saffron cake, French bouillabaisse, Milanese risotto, and Spanish paella. It is also widely employed in Middle Eastern cookery.

Prepared as a tea, saffron is effective as a digestive aid as well as a means of inducing menstruation or perspiration. The herbalist Gerard would write that "moderat use thereof is good for the head, and maketh the sences more quicke and lively, shaketh off heavy and drowsie sleepe, and maketh a man merry."

Saffron is also touted in the Near East as the ultimate aphrodisiac. A book in the Old Testment, the *Song of Solomon,* compares a new bride to a garden that does not lack for saffron. R. Campbell Thompson's *The Assyrian Herbal* (London: Luzac, 1924) notes that the ancient Assyrians frequently used the herb to "set the mood" for the act of love. Saffron's qualities as a perfume and dye (ancient Phoenicians spent their wedding night on sheets colored yellow by the plant) represent key features of the overall experience of intimacy. In addition, the Egyptians and other ancient Mediterranean cultures advocated applying a saffron paste on various parts of the body to ensure sexual arousal during lovemaking. John Heinerman recalls asking a Hartford-based friend, who was known to have a debonair way with women, to try this concoction during one of his amorous sessions. Once they got past the seemingly kinky nature of the request, the man—and his various girlfriends—found the saffron formula to greatly enhance lovemaking. In the man's words, the girls felt that the herb "brought out the best in me instead of the typical 'animal' instincts I've been somewhat notorious for" (p. 410).

BIBLIOGRAPHY

Farrell, Kenneth T. "Saffron." In *Spices, Condiments, and Seasonings*. Westport, CT: AVI, 1985, pp. 195-198.

Garland, Sarah. *The Complete Book of Herbs & Spices*. Pleasantville, NY: Reader's Digest Association, 1993.

Heinerman, John. "Saffron *(Crocus sativus)*." In: *Heinerman's Encyclopedia of Healing Herbs & Spices*. Paramus, NJ: Prentice-Hall, 1996, pp. 408-410.

SAGE

Sage *(Salvia officinalis)* is a hardy, semi-evergreen subshrub indigenous to the northeastern shores of the Mediterranean region. It is cultivated in much of Europe, the Near East, and North America. It rarely grows higher than twenty-four inches, and has long been popular for a variety of culinary, decorative, household, and medicinal uses. The plant prefers full sun and adapts to most types of soil where drainage is efficient. It can be raised from seed, but cuttings produce the best results with variegated and colored-leaf cultivars.

While grown primarily for its foliage, flowers generally appear on sage by early summer. The plant's colorful flowers and leaves render it desirable for herb gardens or in beds and borders with other ornamentals. When crushed, the leaves release a distinctively pungent aroma.

The term "sage advice" appears to have originated in England where sage tea was widely consumed with the belief that it imparted prudence and a strengthened memory. Up until the early modern era, the herb was at the top of the list of household remedies for relieving itching, lowering fevers, and treating nervous headaches. Its value was such that Dutch traders received four pounds of tea from China in exchange for a pound of sage.

Among the many varieties of sage, the most popular include:

- *Icterina* (also known as *Variegata* or golden sage), characterized by leaves with brightly contrasting gold and green coloring
- *Purpurascens* (purple sage), featuring violet-blue flowers and velvety gray-green leaves suffused with purple
- *Tricolor,* whose leaves combine green, cream, white, pink, and purple
- *Albiflora* (white-flowered)
- *Rubriflora* (red-flowered)
- *Aurea* (golden-leaved)

Sage leaves are often used in the construction of fresh herbal wreaths, a practice related to traditional holidays such as May Day as well as modern-day decorative tastes. Like other evergreen herbs that dry naturally in warm air, it combines a pleasing appearance with a lingering scent.

Sage remains best known as a seasoning in cooking. Extremely versatile, it is widely used in cheese and potato dishes, vegetable soups, casseroles, stuffings, sauces, and main courses featuring pork or poultry. It blends extremely well with other herbs like thyme and marjoram.

Sage is valued by advocates of natural cosmetics. The leaves are used in herbal washes aimed at getting rid of dry, flaky skin. They are also combined with rosemary in a rinse for dark hair and dandruff.

Many herbal remedies include sage. A lotion made from the leaves is used for treating acne. The fresh leaves can be applied to treat insect stings. A leaf infusion acts as a stimulant for digestive problems, a mouthwash for gum disorders, and a gargling compound for sore throats. Sage tea can also relieve mouth problems, including toothaches. The essential oil produced from purple sage is often employed in aromatherapy.

BIBLIOGRAPHY

Farrell, Kenneth T. "Sage." In *Spices, Condiments, and Seasonings.* Westport, CT: AVI, 1985, pp. 199-203.

Gardner, Jo Ann. *Living With Herbs.* Woodstock, VT: Countryman Press, 1997.

McHoy, Peter, and Westland, Pamela. *The Herb Bible.* New York: Barnes & Noble Books, 1994.

SARSAPARILLA

Sarsaparilla *(Smilax glauca)* ranked high as a treatment for syphilis in the nineteenth century. Cowboys, sailors, and the like would often ask for it in the form of a refreshing drink (e.g., Ayer's Sarsaparilla) following a trip to a nearby brothel.

Research in recent decades has revealed that the herb provides no discernable benefit against syphilis. It does, however, possess other notable medicinal features; it serves as a diuretic, enhances liver functioning, and assists in the treatment of inflammatory conditions, notably psoriasis and arthritis.

The liver-protective benefits of sarsaparilla have been documented by numerous laboratory experiments involving animals. While not potent enough to serve as an antidote to poisons, the herb binds toxic compounds in the stomach, thereby preventing their movement into the bloodstream. Side effects rarely occur, the one exception being that large doses may upset the stomach.

Perhaps the most popular form of ingesting sarsaparilla is in the form of a decoction. It consists of adding one or two teaspoons of powdered root to a cup of water, which is then brought to a boil and simmered for about fifteen minutes (a three-cup limit is recommended). Because sarsaparilla has an unpleasant aftertaste for many users, herbalists have recommend using a tincture, i.e., one-quarter to one-half teaspoon up to three times daily.

BIBLIOGRAPHY

Heinerman, John. "Sarsaparilla *(Smilax officinalis)*." In: *Heinerman's Encyclopedia of Healing Herbs & Spices*. Paramus, NJ: Prentice-Hall, 1996, p. 423.

"Herbal Healthwatch; Sarsaparilla: Liver Booster." *The Herb Quarterly* (Winter 1999): 8-9.

SASSAFRAS

A smallish tree ranging in height from ten to fifty feet, sassafras *(Sassafras albidum)* possesses grayish to reddish-brown bark known for its rich spicy aroma. Native to the eastern United States, its use is well entrenched in the folkways of Old World settlers.

Brewed as a tea, sassafras bark is used to treat the following conditions: arthritis, dysentery, eczema, fever, gout, herpes, lung problems, psoriasis, rash, rheumatism, and shingles. It is also used to break the smoking habit as well as to slow the flow of milk in nursing mothers. Many rural Americans, particularly in the Appalachian region, believe drinking sassafras tea in the springtime will "thin and purify bad blood."

BIBLIOGRAPHY

Heinerman, John. "Sassfras." In: *Heinerman's Encyclopedia of Healing Herbs & Spices*. Paramus, NJ: Prentice-Hall, 1996, p. 424.

Tucker, Arthur O., Maciarello, Michael J., and Broderick, Cyril E. "File and the Essential Oils of the Leaves, Twigs, and Commercial Root Teas of *Sassafras albidum* (Nutt.) Nees (*Lauraceae*)." In: Charalambous, George (Ed.), *Spices, Herbs and Edible Fungi*. Amsterdam: Elsevier, 1994, pp. 595-604.

SAVORY

Native to the Mediterranean region, savory *(Satureja hortensis)* possesses a distinctive bushy, hairy stem (often taking on a purple hue) that reaches more than one foot in height. Also known as sweet summer savory, its small, oblong-linear leaves are sessile, typically with hairy margins. The two-lipped flowers, either pink or white in color, grow in whorl-like cymes. All parts of this annual are very aromatic.

Highly valued as a culinary spice, savory is cultivated worldwide. The English, who obtained it from the Romans during antiquity, have long included it in their individual herb gardens. The Germans called it the bean herb because it was the perfect complement for green beans, dried beans, and lentils.

Savory has an equally lengthy history as a medicinal herb. Cresentius recommended it "as a purgative, as a remedy in complaints of the liver and lungs and as a bleach for a tanned complexion." Matthioli claimed it stimulated the appetite, promoted digestion, and caused liveliness of the body.

It also has been recognized for centuries as an "herb of love." Banckes' *Herbal* noted, "It is forbidden to use it much in meats since it stirreth him to use lechery." French herbalist Maurice Messegue has frequently recommended it to couples as an aphrodisiac. For this purpose, the herb is typically applied to meat dishes in powdered form. Impotent men and frigid women are advised to drink a decoction of savory and fenugreek as well as applying it to the base of the spine.

BIBLIOGRAPHY

Farrell, Kenneth T. "Savory, Sweet Summer." In: *Spices, Condiments, and Seasonings*. Westport, CT: AVI, 1985, pp. 204-207.

Gardner, Jo Ann. *Living with Herbs*. Woodstock, VT: Countryman Press, 1997.

Heinerman, John. "Savory *(Satureja hortensis)*." In: *Heinerman's Encyclopedia of Healing Herbs & Spices*. Paramus, NJ: Prentice-Hall, 1996, pp. 424-425.

SAW PALMETTO

Saw palmetto *(Serenoa repens)* is a type of palm tree, also referred to as the dwarf palm. Its active medicinal ingredient is found in the oily compounds within its berries. Saw palmetto dietary supplements in the United States are promoted primarily to improve urinary flow and reduce the frequency and urgency or urination in males with prostate enlargement. The herb is also used to treat bladder inflammation (cystitis), chronic bronchitis, laryngitis, asthma-associated nasal inflammation, and other conditions.

Scientists believe the fatty acids and sterols present within saw palmetto inhibit the actions of testosterone on the prostate that cause prostate enlargement and interference with urinary flow. At least 85 percent of the weight of clinical-quality saw palmetto products is composed of these constituents.

The majority of health supplements in the marketplace are composed of an extract from the berries; however, crushed berry products are also available. Because saw palmetto and other dietary supplements are not considered drugs in the United States, they are not legally required to be tested for quality by any governmental or independent agency. Given the proliferation of products based on a saw palmetto extract, the consumer faces a confusing array of choices. Reputable, high-profile manufacturers currently marketing saw palmetto in extract form include Amway, Bayer, CVS, Celestial Seasonings, Centrum Herbals, Enzymatic Therapy, GNC, MotherNature. com, Natrol, Nature's Way, NaturPharma, Pharmassure, PhytoPharmica, Puritans' Pride, Quanterra, Shaklee, Sundown Herbals, Sunsource, and Walgreens. Discount Natural Herbs, a leading Internet retailer, included saw palmetto in its "Top Ten Selling Herbs" as of mid-2001.

In an attempt to establish quality standards and better inform consumers, ConsumerLab.com has instituted a testing program. Furthermore, the agency is licensing manufacturers and distributors to apply its flask-shaped Seal of Approved Quality for saw palmetto on labels and in advertising for products that have passed testing.

Saw palmetto use is expected to maintain an upward curve based upon Overmyer's dissemination of findings of the first American randomized clinical trial in which the herb proved beneficial in reducing swelling of prostate tissues in patients with benign prostatic hyperplasia (BPH). The fact that the saw palmetto extract (SPE) produced this benefit without affecting testosterone or dihydrotestosterone levels means it works by some nonhormonal mechanism. In other words, SPE appears to be a viable alternative to conventional drug therapy for BPH, which frequently causes undesirable hormonal side effects (e.g., reduced sex drive and performance). Furthermore, SPE does not affect the level of prostate specific antigen (PSA) in the blood. In contrast, conventional therapy interferes with PSA testing, which is used to detect prostate cancer.

BIBLIOGRAPHY

ConsumerLab.com. "Product Review: Saw Palmetto," www.consumerlab.com/results/sawpalmetto.html LLC, 2000. 4p.

Griggs, Barbara. *Green Pharmacy: The History and Evolution of Western Herbal Medicine*. Rochester, VT: Healing Arts, 1997.

McCaleb, Rob. "Saw Palmetto Shrinks Prostate Tissues." *Herb World News Online*. Boulder, CO: Herb Research Foundation, 2000. 2p. www.herbs.org/current/sawshrink.html. Expanded article available online for $20.

Overmyer, M. "Saw Palmetto Shown To Shrink Prostatic Epithelium." *Urology Times 27*(6) (1999): 1, 42.

SELF-HEAL

Self-heal *(Prunella vulgaris)* is often found as a "weed" in neighborhood lawns and gardens across Europe and North America. In many instances, home dwellers will allow it to mature, or even cultivate it, due to its attractive purple flowers. What many of them don't realize is that the herb has a wide array of medicinal uses.

While the plant, also known as heal-all, remains relatively small (two to three inches in height) in open areas, it tends to attain a larger size in sheltered settings. The main stem tends to be grooved with purple highlights at its base. The aerial parts—the portions employed as herbal medicine—include oblong leaves and flowers that form a spike at the top of the plant. Self-heal blooms in midsummer, the ideal time for collecting its usable parts.

The herb's active constituents consist of flavonoids, vitamins (A, B, C, and K), fatty acids, a volatile oil, and a bitter compound. In Chinese medicine, the flower spikes are utilized to cool "liver fire," the symptoms of which include anger (or temper tantrums), hypertension, constipation and dark yellow urine, nosebleeds, red eyes, and a red tongue with a yellow coat.

In the West, self-heal is highly prized for its antibacterial, astringent, and diuretic properties. Internally, a tea extract is used as a spring tonic and as a gargle for sore or ulcerated throats as well as to slow menstrual bleeding, to curb diarrhea, and to stop bleeding within the body, most notably the gastrointestinal tract. Externally, the fresh leaves can be made into a poultice to stop bleeding. Furthermore, healing of these wounds (as well as hemorrhoids) can be facilitated through application of a self-heal ointment.

One caveat associated with use of the herb: it can potentially undermine the intended benefits of prescription blood thinners such as Coumadin, dipyridamole, heparin, Persantine, Plavix, and warfarin.

BIBLIOGRAPHY

Grieve, Maud. *A Modern Herbal, Volume II*. New York: Dover, 1981, pp. 731-733.

Langer, Robin E. "Herb-of-the-Month: Self-Heal or Heal-All *(Prunella vulgaris)*—April 2000." *Snowbound Herbals* (April 2000): 2p. www.sbherbals.com.

McHoy, Peter, and Westland, Pamela. *The Herb Bible*. New York: Barnes & Noble Books, 1994.

SHAKER HERBS

Development of the Shakers' Herb Business

In 1747, Jane and James Wardley, former Quakers living in Manchester, England, established a new religious society based on such beliefs as the open confession of sins and taking up testimony against sin. By 1774, Ann Lee and a small band of followers left the Wardleys in England and sailed to New York to win new converts. They eventually called themselves Believers in Christ's Second Appearing, but, more popularly, the sect became known as Shakers, because of their demonstrated religious fervor. Highlights of early Shaker history included:

1787	The Shaker community at Watervliet, New York, is formally established and a seed industry initiated here in 1790.
1794	A Shaker herb business is started in New Lebanon, New York. By 1795, the villages at New Lebanon, Hancock (Pittsfield, Massachusetts), Canterbury, New Hampshire, and Enfield, Connecticut, have begun similar seed businesses.
1801	A number of Maine communities start selling seeds.
1802	An Enfield, Connecticut, Shaker group establishes a seed concern.
1812	New Lebanon Shakers begin an herb garden.
1816	Thomas Corbett establishes an herb garden at Canterbury while South Union, Kentucky, members take their first flatboat trip down the river to New Orleans to find new markets for seeds and other products.

Around 1820, the Shakers began to prepare roots and herbs for sale, by drying, pressing, and papering them in compact form. Although among the first to establish herb businesses in the United States, Shaker endeavors began on a modest scale. They had no routine for finding the herbs they needed, gathering the plants in surrounding fields and woods wherever they could find them. Children often collected the herbs, and the Shakers also enlisted the help of local Indians who were experts at identifying indigenous plants. When necessary, they traveled great distances to gather the plants. The Shakers gradually developed procedures for collecting the wild herbs, and a form of quality control helped make their products desirable.

In 1830, the Watervliet Believers issued the first dated Shaker herb catalog, offering 128 herbs for sale, along with pills, syrups, and two medicinal waters. Three years later, the Union Village Shakers established an herb garden. In 1835, the Shaker village at Canterbury issued its first herb catalog, listing 180 herbs for sale. Also in 1935, the communities of New Lebanon, New York, and Enfield, Connecticut, printed the *Gardener's Manual.*

The Shaker physicians were excellent botanists who selected the correct plants for their medical needs; they were knowledgeable in many factors that affected the potency of their herbs, including soil identification, weather, and pest conditions. These doctors determined not only what was to be harvested but also when it was to be done. The Shakers collected one variety of a plant at a time, at the peak of its growth, and only the part that was needed (leaves, flowers, roots, seeds, etc.). A single plant could have several different medical purposes. The bark of the hemlock tree, for example, was used to treat gangrene, diarrhea, leucorrhea, and a prolapsed uterus; the oil from the leaves was added to liniments; and the gum was saved for making plasters. In the fall, the heavy roots were gathered in baskets and brought indoors to be cleaned, split, and thoroughly dried with artificial heat. The more fragile parts of the herbs were collected on tow sheets, usually fifteen feet square. Flowers, the most valuable part of the plant, were taken at their first opening and were quickly dried to retain their colors.

The quality of their herbs earned the Shakers a reputation as fine horticulturists. As the demand for their products increased, they expanded their sales beyond the local area. One of their most important targeted markets was the medical community, whose members apparently were dissatisfied with their imported plants.

Out of necessity, the Shakers began planting gardens to ensure a steady supply of nonnative or scarce plants. In order to meet the increasing demand for their herbs and grow certain plants under carefully controlled conditions, they created scientifically organized "physic" or botanical gardens. These plots of land required relatively little acreage since many herbs continued to be harvested in the wild. At the peak of their business, the Shakers had about 150 acres of herbs under cultivation. The rest of their fields were devoted to such crops as hay and wheat, or pasturage. The first known mention of the Shakers growing herbs in an organized fashion is recorded in a diary by a New Lebanon Brother called Jethro, who wrote that in June 1812 "a violent storm did much damage in herb garden."

The Shakers' physicians were initially responsible for these gardens. They transplanted the herbs from the wild and grew them from seed. Their acreage was kept separate from the Shakers' other plots that supplied the communities' vegetables and seed businesses. To identify their plants and ensure a steady and pure supply for the market, the doctors referred to botanical books from outside the Shaker community, such as *A Manual of Botany for the Northern States and Medical Flora*.

From community to community, the physicians communicated freely with one another, often writing and visiting their sister societies and ordering herbs from them. By negotiating, the doctors could guarantee high-quality herbs at a lower cost. They bargained for the best prices, offered discounts, and gave one another preference in filling orders when it was economically feasible in order to keep their fees competitive with the World's (non-Shaker community).

By 1831, the New Lebanon Shakers were shipping boxes of herbs around the world. In the next three years, they were harvesting 4,000 pounds of roots and herbs annually. At the same time, they published their first catalog offering 137 herbs for sale. In 1849 they sold 16,500 pounds of herbs and 587 pounds of extracts. The following year they had fifty acres devoted to herb gardening and had sold 21,000 pounds of herbs and 7,000 pounds of extracts. By the 1830s, the South Union, Kentucky, Shakers had a profitable seed business (established in 1821) and a distillery, herb press, and buildings in which they processed herbs primarily for their own use. An herb catalog from this community has not been found; however, whatever herb business they did have was greatly accelerated during the Civil War because of the increased demand for medicines.

Community Sharing

In many bishoprics, particularly where there were villages that were located close together and had the same clients, one community would raise herbs and another, seeds. If a community raised both, one would be for trade and the other would be for home use. This mutual support between communities is seen in the bishopric shared by the Shirley and Harvard communities in Massachusetts. Shirley depended on income from the sale of its vegetable and herb seeds. In 1805, the village was selling caraway, lavender, parsley, sage, saffron, and summer savory seeds. Ten years later, it was also selling balm, burnet, and fennel seeds. While the actual date of the establishment of Harvard's gardens has not been recorded, it is known that this community had prepared herbs for the Boston market by 1820. The business grew slowly, but eventually the community issued eleven bound catalogs, the first in 1845. Shirley helped out its sister community by gathering green herbs and planing the wood for Harvard's herb boxes.

The fact that the Shakers could regularly depend on one another for high-quality herbs, even during times of crop failures and unexpected new markets, contributed to their reputation as reliable dealers. Their communal lifestyle made it possible for them to sell in large volume to the burgeoning pharmaceutical market. They had enough members so that growing, harvesting, processing, and advertising their herbs and medicines could be done efficiently at a lower cost, and they could sell their products at competitive prices.

In 1848, the Enfield, Connecticut, Shaker community published its *Extract of English Valerian* catalog to meet public demand for that particular herb. Three years later, Canterbury issued the first *Shakers' Manual,* which advertised both its Corbett's Sarsaparilla Syrup and Enfield's Valerian Extract. In 1873, the Shaker herb catalogs were compiled as *Druggist's Hand-Book of Pure Botanic Preparations, etc.* Sold by Society of Shakers, Mount Lebanon, Columbia County, New York, it listed 304 herbs with their Latin names, the medical properties ascribed by the Shakers, and a synopsis of the various diseases for which they were used. This herbal compendium survived, but the Shakers did not; in 1965, the Lead Ministry voted to close the covenant membership of the Shaker societies. Today, only a few members remain and with their deaths, the society will become extinct.

BIBLIOGRAPHY

Beale, Galen, and Boswell, Mary R. *The Earth Shall Blossom: Shaker Herbs and Gardening*. Woodstock, VT: Countryman Press, 1991.

Buchanan, Rita. *The Shaker Herb and Garden Book.* Photography by Paul Rocheleau. Boston: Houghton Mifflin, 1996.

Buchanan, Rita. "Shaker Herbs." *The Herb Companion* (February/March 1997): 52-57.

Miller, A.B.W. *Shaker Medicinal Herbs: A Compendium of History, Lore, and Uses*. Pownal, VT: Storey Books, 1998.

SMARTWEED

Smartweed *(Polygonum hydropiper)* is named for the pronounced peppery taste of its narrow leaves and reddish jointed stems. During summer months tiny, greenish-white (typically red-lipped) blossoms appear on long, slender, dropping spikes. The annual thrives across the United States with the exception of southern Georgia and the Florida peninsula. It is a close relation of knotgrass *(Polygonum aviculare)*, whose astringent qualities are useful in treating diarrhea.

The discovery of smartweed seeds at Paleo-Indian sites located within the Ohio and Illinois River valleys indicates that this herb was important to many prehistoric cultures. The concentration of these seeds in garbage pits and campfires has caused archeologists to surmise that smartweed was popular as a flavoring agent given its hot, biting taste at a time when salt and pepper were virtually unknown to area inhabitants.

Paleo-Indians also appear to have utilized the herb for medicinal purposes. A tea brewed from cut stems and leaves has been demonstrated to effectively treat hypertension and hasten blood coagulation in body wounds.

BIBLIOGRAPHY

Griggs, Barbara. *Green Pharmacy: The History and Evolution of Western Herbal Medicine*. Rochester, VT: Healing Arts, 1997.

Heinerman, John. "Smartweed *(Polygonum hydropiper)*." In: *Heinerman's Encyclopedia of Healing Herbs & Spices*. Paramus, NJ: Prentice-Hall, 1996, pp. 430-431.

SORREL

A perennial originating in grassy areas of Britain, Europe, and Asia, sorrel *(Rumex acetosa)* is widely cultivated in the United States. The flowering stems reach three feet in height and are punctuated by broad, arrow-shaped leaves and spikes of tight, clustered, red flowers beginning in early summer. Propagation is by seed and division. It is necessary to divide plants as well as cut down flowering stems to provide an ongoing harvest of succulent leaves.

Sorrel leaves have many culinary uses, most notably in salads (or in place of spinach), soups, and sauces. The leaves and flowering tops are utilized in producing greenish-yellow dyes.

Sorrel is known for its cleansing properties; it is widely taken for fevers as well as bladder, liver, and kidney problems. Sorrel tea can alleviate the pain and the need for removing gall and kidney stones. Swishing the tea inside the mouth treats mouth sores; gargling clears up a soar throat or throat ulcers. Juice from the leaves can be administered as a poultice for skin troubles. In addition, a concentrated infusion removes stains from linen, wicker, and silver.

BIBLIOGRAPHY

Gardner, Jo Ann. "Sorrel (Garden Sorrel)." In: *Living with Herbs*. Woodstock, VT: Countryman Press, 1997, pp. 240-243.

Garland, Sarah. *The Complete Book of Herbs & Spices*. Pleasantville, NY: Reader's Digest Association, 1993.

Heinerman, John. "Sorrel *(Rumex acetosa)*." In: *Heinerman's Encyclopedia of Healing Herbs & Spices*. Paramus, NJ: Prentice-Hall, 1996, p. 432.

ST.-JOHN'S-WORT

St.-John's-wort *(Hypericum perforatum)* is a perennial herb ranging from one to three feet in height. It belongs to the genus *Hypericum,* which is composed of over 400 species of trees, shrubs, and herbs. The designation *perforatum* comes from the translucent dots borne by the leaves. The plant's popular name was bestowed by the early Christians in honor of John the Baptist because they thought the flowers released their blood-red oil on August 29, the anniversary of the disciple's beheading.

St.-John's-wort is native to Europe where it continues to thrive except in the extreme north. First-century Greek physicians Galen and Dioscorides employed it as a diuretic, wound healer, and treatment of menstrual disorders. It enjoyed considerable renown during the Middle Ages as a talisman to protect against demons.

The superstitions associated with St.-John's-wort led many physicians to dismiss it as a folk medicine by the mid-nineteenth century. Eclectic medical practitioners in the United States, however, continued to utilize it for healing, most notably lacerations involving nerve damage, in addition to diuretic, astringent, and sedative applications. By the early twentieth century, the plant was widely used in Europe for neuroses, general restlessness, insomnia, and mental or emotional disorders caused by intellectual exertion.

Beginning in the 1990s, the United States imported a plethora of European home remedies and pharmacy preparations which employed St.-John's-wort. Among the more popular applications are:

- Vegetable oil extracts used externally for hemorrhoids
- A tea derived from the plant that serves as a mild nerve tonic for anxiety, depression, and insomnia, and as a diuretic and treatment for gastritis
- Both alcohol tinctures and internal dosages that have been prescribed as antibacterials

In the rush to utilize St.-John's-wort, however, researchers have advised the public that it is unsuitable for treating severe depression. Photodermatitis—a condition afflicting light-skinned livestock who have eaten the plant (the increased sensitivity to light causes swelling, blindness, and often death from starvation)—is considered a potential problem for fair-skinned persons exposed to the sun while taking St-John's-wort internally. Furthermore, the herb should not be taken simultaneously with conventional antidepressant drugs, particularly monoamine oxidase (MAO) inhibitors such as Prozac. Nevertheless, many herb advocates have continued to recommend St.-John's-wort as an antidepressant (see articles by McCaleb and Morien). Findings published in the prestigious *New England Journal of Medicine* that the herb is not effective in fighting severe depression—a major news story in many mass media outlets on April 17, 2001—represent yet another chapter in this ongoing controversy.

BIBLIOGRAPHY

Awang, D.V.C. "St.-John's-wort." *Canadian Pharmaceutical Journal 124* (1991): 33-35.

Ernest, E. "St.-John's-wort, An Anti-Depressant? A Systematic, Criteria-Based Review." *Phytomedicine 2*(1) (1995): 67-71.

Foster, Steven. *Herbs for Your Health*. Loveland, CO: Interweave Press, 1996.

Foster, Steven. "St.-John's-wort: Past and Present." *The Herb Companion* (February/March 1997): 72-73.

Hahn, G. "*Hypericum perforatum* (St.-John's-wort): A Medicinal Herb Used in Antiquity and Still of Interest Today." *The Journal of Naturopathic Medicine 3*(1) (1992): 94-96.

Harter, G., and Sommer, H. "Treatment of Mild/Moderate Depressions with Hypericum." *Phytomedicine 1*(1) (1994): 3-8.

Hobbs, C. "St.-John's-wort." *HerbalGram 18/19* (1988/1989): 24-33.

Leigh, Evelyn. "Efficacy of St. John's Wort Extract Standardized to Hyperforin." *Herb World News Online*. Boulder, CO: Herb Research Foundation, 2000. 2p. www.herbs.org/current.

McCaleb, Rob. "St. John's Wort Equivalent to World's Best Selling Antidepressant." *HerbalGram 49* (2000): 26-27.

Morien, Krista. "Safety of St. John's Wort vs. Antidepressant Drugs." *HerbalGram 49* (2000): 24-25.

Morien, Krista. "Three Concentrations of St. John's Wort Effective Against Depression." *HerbalGram 49* (2000): 25-26.

Purkh, Karta. "Celebrity Herbs, Covert Uses." *The Herb Quarterly* (Fall 2000): 21-25.

Reichert, R. "St.-John's-Wort Extract as a Tricyclic Medication Substitute for Mild to Moderate Depression." *Quarterly Review of Natural Medicine* (Winter 1995): 275-278.

STEVIA

It is not surprising that the virtues of stevia *(Stevia rebaudiana)* remain undiscovered to most herb users within the United States. Nothing is particularly distinctive about the plant's outward appearance; it reaches two to three feet in height, and possesses toothed, opposite leaves, hairy stems, and clusters of tubular white blossoms.

Up to the 1970s, stevia was virtually unknown outside its native region of Paraguay and Brazil. The plant's undiluted extracts are roughly 300 times sweeter than sugar. For centuries, it was used to sweeten tea and medicines by the Guarani Indians. The herb has been marketed in the rest of the population in these two nations, primarily as a sugar substitute for diabetes. Since the early 1970s, stevia has been extremely popular in Brazil and Japan as a natural noncaloric sugar substitute in soft drinks, chewing gum, soy sauce, pickles, and many additional food products. It was also being widely used in much of Asia by the 1990s.

Stevia's flavor differs considerably from that of sugar; its taste more closely approximates the nectar of wild honeysuckle, albeit much sweeter and slightly stronger. The herb can effectively sweeten virtually any food dish. In baking, best results are obtained when the leaves are dried and finely ground with a mortar and pestle. Dried stevia leaves are sweeter than fresh ones and maintain their flavor for months. Stevia can also be used in liquid concentrate form.

The discovery in the early 1990s that stevioside, the compound behind the plant's sweetness, also reduces blood sugar levels has spurred experiments regarding its medicinal value. Marketing was held up in the United States for many years while scientific and regulatory organizations tested its potential as a commercial sweetener. Nevertheless, stevia's popularity increased rapidly during the 1980s and early 1990s as its reputation continued to grow. As a result of the Health Freedom Act, passed in September 1995, stevia leaves and extracts are allowable for import and commerce in the United States if labeled as a dairy supplement and used as an ingredient in a dietary supplement. Sweetvia stevia extract and Sweevia single-serving packets represent products marketed to consumers seeking natural foods, weight-conscious individuals, those suffering from blood sugar disorders, and the yeast intolerant.

Domestic herb suppliers carry the herb in powdered or bulk form, and some health food retailers offer it as an extract. Nurseries sell seeds, rooted cuttings, and small plants; supplies are generally limited and so are quickly depleted. The FDA has yet to approve the product for use as a food additive due to lack of expensive testing and suspected lobbying pressure by companies producing synthetic sugar substitutes.

Limited availability has rendered private cultivation an attractive alternative. The herb can be grown as a tender perennial in most of the United States. It winters over well in the southwest; where winters are cold, it should be potted and kept indoors. Successful cultivation depends upon early spring planting, a regular watering schedule, and frequent pruning. Once established, the plants are relatively undemanding and disease-free.

BIBLIOGRAPHY

"At-Stevia." Valley Forge, PA: Home Sweet Home, n.d. www.at-stevia-recipes-book.com.

"Stevia Has Been Given FDA Approval." 1pagefx, January 7, 1996. 4p. www.mcs.net/~joyce.

Thomas, Uwanna D. "An Herb To Know." *The Herb Companion* (December/January 1991/1992): 18-19.

STINGING NETTLE

Common in temperate climates, Africa, Asia, Australia, Europe, and the Americas, stinging nettle *(Urtica dioica)* is a perennial reaching about three feet in height. Its distinctive dark-green, toothed, heart-shaped leaves are covered with stinging hairs. The sting from some of the more than 500 species of nettles requires several months of recuperation; on occasion, it can cause death. The agents behind the sting are formic acid and histamine. Antidotes include yellow dock (or curly dock), rosemary, mint, sage, the juice of the plant itself, and commercial toothpaste products.

After midsummer, hanging clusters of green or reddish flowers are pollinated by the wind. The plant quickly spreads by seed and aggressive creeping roots. For medicinal use, the entire nettle should be gathered before flowering and hung for drying in a shady, airy location.

Although difficult to harvest, the herb provides many benefits. During the expansion into Europe and the British Isles, the Romans brought along nettles, believing they would have to beat themselves with the plant in order to keep warm. Arthritis and rheumatism were also treated in this manner for many centuries. Before flax and hemp came along, the stalk fibers were employed in making cloth, fishing nets, rope, and similarly sewn products.

Today, the young leaves are a popular vegetable, both for their flavor and as a source of vitamins and minerals. Nettles contain acetylcholine, calcium, chlorine, chlorophyll, formic acid, glucoquinones, histamine, iodine, iron, magnesium, potassium, serotonin, silicon, sulfur, tannin, and vitamins A, B, C, and K. The abundance of iron and vitamin C are especially useful in treating anemia in that the latter enhances iron absorption from the GI tract. Nettle tea is commonly combined with other drinks and juice concentrates to improve nutritional benefits. The herb is also used in soups, puddings, and the making of beer.

The aerial portions of the plant are useful as an astringent, diuretic, expectorant, hemostatic, circulatory stimulant, and nutritive tonic, in addition to improving milk flow in nursing, lowering blood sugar levels, treating gout and arthritis, and preventing scurvy. When boiled, they produce a green dye.

The root produces a yellow dye when boiled and mixed with salt. Combined with other herbs such as pygeum and saw palmetto, it can relieve symptoms of benign prostatic hypertrophy. It is also utilized as a conditioner to fight dandruff and hair loss.

Nettles aid in mitigating the symptoms of hay fever and other forms of sinus congestion. The herb effectively treats virtually any condition relieved by improved mineral balance, including goiter, high blood pressure, malabsorption syndrome, muscle cramps, and osteoporosis.

Nettles can be taken in a variety of forms. In addition to serving fresh, they are used as a tea, tincture, compress, and in ointment or powdered form. The root is often tinctured or simmered in water to create a decoction.

While producing no known side effects, complications can arise from employing nettles with a number of medications. For instance, their vitamin K content may inhibit the effectiveness of prescription drugs designed to thin the blood, lower blood pressure, or have a diuretic effect.

Despite this caveat, nettles were one of most widely used herbs worldwide at the outset of the twenty-first century. The Lacey, Washington-based Internet retailer, Discount Natural Herbs included the plant in its "Top Ten Selling Herbs" listing through mid-2001.

BIBLIOGRAPHY

Garland, Sarah. *The Complete Book of Herbs & Spices*. Pleasantville, NY: Reader's Digest Association, 1993.

Grieve, Maud. *A Modern Herbal, Volume II.* New York: Dover Publications, 1981, pp. 574-579.

Langer, Robin E. "Herb-of-the-Month: Stinging Nettle *(Urtica Dioica)*—December 1999." *Snowbound Herbals* (December 1999): 2p. www.sbherbals.com.

SUNFLOWER

The sunflower *(Helianthus annuus)* is perhaps best known for its large flower heads (up to forty centimeters wide), which are composed of yellow ray florets and purplish-brown, tubular disc florets. The tall annual possesses an erect, often branched hairy stem punctuated by large, chordate leaves (opposite at lower end and alternate in upper reaches). The fruit, a faintly flattened achene, is often streaked with white and black.

The common and generic names were allegedly derived from the plant's inclination to face the sun's direct rays by day. However, the similarity of the flower heads to the sun represents another possible name source.

Native to western North America, the herb was cultivated by American Indians prior to 1000 B.C. Introduced to Europe in the sixteenth century, it became a major food plant when it reached Russia. It remains popular worldwide today, being commercially grown for oil, fodder, and decorative purposes.

The seeds—whose fatty oil contains glycerides of unsaturated linolenic and oleic acids and saturated palmitic and arachic acids—are the most useful portion of the sunflower. They are contained in salves, plasters, and liniments that treat rheumatic pain. The seeds are widely used in foodstuffs (e.g., salad, margarine); roasted, they are eaten plain, ground into flour, and substituted for coffee. They also have value as an ingredient in soaps and lubricants. In homeopathy, a tincture can be administered internally to relieve constipation and externally for cuts and bruises.

The dried flower heads are employed medicinally in many countries where they are recognized as having anti-inflammatory, antidiarrheal, carminative, and diuretic properties. In the past, the leaves and flowers were utilized to treat malaria.

BIBLIOGRAPHY

Bunney, Sarah (Ed.). *The Illustrated Encyclopedia of Herbs; Their Medicinal and Culinary Uses*. New York: Barnes & Noble Books, 1996. (Based on *The Illustrated Book of Herbs*, by Jiri Stodola and Jan Volak [Aventinum Nakladatelstvi, 1990].)

Garland, Sarah. *The Complete Book of Herbs & Spices*. Pleasantville, NY: Reader's Digest Association, 1993.

T

TEA

According to Chinese legend, tea was discovered by the Emperor Shen Nong in about 2700 B.C. when a gust of wind blew tea leaves into a kettle of boiling water. Indian legend, however, credits Siddhartha Gautama, founder of Buddhism, with the divine creation of the drink.

Allegedly the second most widely consumed beverage worldwide (behind water), tea *(Camellia sinensis)* has been cultivated at least as far back as the fourth century A.D. It was confined largely to Asia until the late sixteenth century, when European explorers established trading ties with the Orient. From that point onward, tea was a highly prized commodity in virtually all cultures.

Iced tea is said to have first been served at the 1904 World's Fair in St. Louis. In the midst of an oppressive heat wave, a vendor whose hot tea was selling poorly poured the beverage over ice for customers. The new concoction was an immediate hit; its fame quickly spread across the nation. By the late 1990s, Americans were consuming more than thirty-three billion glasses of iced tea annually.

In addition to its enticing flavor, scientific research has linked healthful benefits to black tea consumption. A large-scale Dutch population study revealed that the drink may decrease the risk of developing atherosclerosis, a hardening and narrowing of the coronary arteries that contributes to heart attack, stroke, and other cardiovascular diseases. This study reinforced earlier research in suggesting that tea flavonoids may protect against heart disease; however, most earlier studies had focused on green tea, due to its higher flavonoid content.

Green tea experienced a spectacular upsurge in popularity in North America during the 1990s when studies revealed that those drinking the Japanese beverage were less likely to develop cancer. The difference between the two types of tea appears minimal; green tea is based on dried leaves, whereas black tea involves fermenting the dried leaves, resulting in a sharper, richer, more tannic flavor. Fermenting does not appear to rob tea of its antioxidants; animal research indicates that black tea also possesses significant anticancer benefits. Nevertheless, marketing to health-conscious consumers in recent years has focused on the benefits accruing from green tea consumption such as brightening teeth and slowing down the aging process. One radio infomercial aired in the spring of 2001 noted that such potential benefits required the consumption of some thirty cups of green tea a day; hence, the advertising company was offering a special liquid concentrate geared to providing comparable results.

Aware of the increased demand for lower-caffeine alternatives to black tea in the past decade, all major U.S. tea producers now market herbal tea blends; many of these are designed to be served over ice. Many tea drinkers, however, have opted to make their own blends employing fresh or dried herbs obtained from domestic gardens, health food outlets, and even mainstream supermarkets.

While the variations of herbal tea are virtually unlimited, the following list—grouped by broad flavor categories—reflects the ingredients preferred as accents by present-day tea afficionados:

- Spicy favors: allspice, aniseed, cinnamon, cinnamon basil, cloves.
- Fruity flavors: chamomile, citrus zest, gingerroot, lemon basil, lemon thyme, lemon verbena, lemongrass, pineapple sage, raspberry leaves.
- Floral flavors: hibiscus, jasmine, lavender, rose hips, rose petals.
- Minty flavors: bee balm, catnip, peppermint, spearmint.
- Herbal flavors: hyssop, marjoram, parsley, rosemary, sage, savory, yarrow.

BIBLIOGRAPHY

Buchanan, Rita. "'Real' Tea." *The Herb Companion* (December/January 1991/1992): 26.

Geleijinse, Johanna M., and Launert, Lenore S. "Tea Flavonoids May Protect Against Atherosclerosis: The Rotterdam Study." *Archives of Internal Medicine 159* (1999): 2170-2174.

"Green Tea and Breast Cancer: The Latest News." *The Herb Quarterly* (Winter 1999): 6.

Henderson, Robert K. "Labrador Tea: The Grandmother of Herbal Teas." *The Herb Companion* (February/March 1997): 58-61.

Langsner, Louise. "Herbs for a Tea Garden." *The Herb Companion* (December/January 1991/1992): 27-32.

Leigh, Evelyn. "Black Tea May Protect Heart Health." *HerbalGram 48* (2000): 22-23. Also available at www.AllHerb.com.

Scano, Audrey. "Herb Tea on Ice." *The Herb Companion* (April/May 1997): 42-45.

THYME

This is one herb whose idiosyncratic pronunciation should be ingrained in the consciousness of most inhabitants of the Western world. Simon and Garfunkel's "Parsley, Sage, Rosemary, and Thyme"—the duo's 1966 classic album containing the familiar "Scarborough Fair/Canticle," heard on the soundtrack of the landmark film, *The Graduate*—deserves much of the credit for this fact.

Thyme refers to the general name for the numerous herbs of the *Thymus* species, all of which are small, perennials native to Europe and Asia. The principal type is garden thyme *(T. vulgaris);* although the chief variants are English, French, and German—which differ by leaf shape, leaf color, and essential oil composition—it is cultivated throughout Europe and the western United States. It is a low-growing, woody shrub with leaves spanning the gray-to-green spectrum and white, pink, or purple flowers.

Ancient Greeks considered it symbolic of courage. They valued the honey obtained from the thyme blossoms that proliferated around Athens; to smell of thyme was the ultimate Greek compliment. It was widely employed in antiquity for incantations and charms. It was also an ingredient in the Egyptian embalming process.

While used primarily for ornamental purposes, thyme also possesses considerable medicinal value. According to the ancient physician Dioscorides, thyme ingested with meat improves poor eyesight. He also recommended mixing it with honey for asthma, expelling worms, killing lice, and driving out the phlegmy matter from the thorax.

Maurice Messegue, considered by John Heinerman to be Europe's most renowned folk healer, had the following to say about the herb's antiseptic qualities:

> [Thyme] contains thymol and its smell destroys viruses and bacteria in the atmosphere as it destroys infectious germs in the body. I do not know any infection that cannot be mitigated if treated with this precious herb. It is an excellent weapon against epidemics and much cheaper than other means of controlling them. From boils to typhoid and whitlows to tuberculosis, it is excellent beyond compare! (Heinerman, 1996, p. 446)

Messegue's catalog of conditions that could be effectively treated by thyme includes:

- Sore throat, bad breath, tooth decay, cold sores, common cold, influenza, fever, and allergies by means of internal (gargling and drinking) applications
- Lung congestion (asthma, bronchitis, etc.), bumps, bruises, and aching joints and muscles by means of external (hot compress and lotion) applications
- Nail fungus, athlete's foot, yeast infection of the vagina, and poor blood circulation by means of soaking or douching
- Wounds, burns, and sore eyes by means of washing
- Sores, boils, flu, poor appetite, and upset stomach by means of various food recipes
- Tooth and gum diseases by means of an herbal toothpaste composed of fresh thyme and brandy

Thyme is combined with ghee (clarified butter) and melted beeswax to make a homemade salve effective in healing acne, bruises, cuts, and rashes on the face and neck. The salve is also useful for treating burns, sores, and wounds all over the body.

In addition to its antiviral properties, thymol assists in relaxing tense muscles and tight blood vessels. A tea made from fresh or dried thyme can help relieve migraine headaches or stomach cramps. This tea (Listerine antiseptic is an acceptable substitute) can also be applied as a compress to treat these conditions. Both Listerine and the popular Mentholatum Deep Heat Rub contain significant amounts of both thymol and eucalyptol from eucalyptus.

Thyme—as well as peppermint, rosemary, sage, and savory—is widely employed to purify water in Mexico, Spain, Portugal, Greece, Italy, and Russia. A handful of cut thyme combined with a quart of water—which is then boiled, allowed to simmer, and strained—produces safe drinking water unlikely to cause diarrhea, fever, and other unhealthy reactions.

In recognition of thyme's growing culinary appeal, McCormick's *Flavor Trends 2000* included the herb in its "Top 10 Flavors for 2000."

BIBLIOGRAPHY

Farrell, Kenneth T. "Thyme." In: *Spices, Condiments, and Seasonings*. Westport, CT: AVI, 1985, pp. 217-221.

Gardner, Jo Ann. "Thymes." In: *Living with Herbs*. Woodstock, VT: Countryman Press, 1997, pp. 257-260.

Heinerman, John. "Thyme *(Thymus vulgaris)*." In: *Heinerman's Encyclopedia of Healing Herbs & Spices*. Paramus, NJ: Prentice-Hall, 1996, pp. 445-449.

TROUT LAKE FARM

Trout Lake Farm is located in the Cascade Mountains of southern Washington near the Columbia River. By the early 1980s, the enterprise included 250 acres devoted to organic farming, with herbs comprising the bulk of the produce. Herbs grown include anise, hyssop, catnip, comfrey, echinacea, lemon thyme, mints, and raspberry leaves.

Trout Lake Farm's herbs are purchased by so-called "conscientious distributors," primarily for beverages, health care products, and pet products. Owner Lon Johnson states, "We think people should have access to wholesome, organically grown herbs. Especially those herbs used for beverages and medicinal uses should be grown without toxic sprays and fumigants. Unfortunately, this is not the case with most imported and domestically grown herbs" (Meares, p. 1).

The farm includes a large drying barn as well as milling equipment similar to what exists in old grain mills. Cut and sifted herbs are sold to distributors both packaged and in bulk. With herbal product use on the rise, Trout Lake is constantly exploring new marketing possibilities. For example, Johnson has considered the use of mint stems for kitty litter.

BIBLIOGRAPHY

Meares, Portia. "Trout Lake Farm: Organic Herb Growers." *The Business of Herbs II*(2) (May/June 1984): 1-3.

TURMERIC

Part of the ginger family, turmeric *(Curcuma domestica)* is cultivated in China, India, Java, and Peru. The leaves are approximately twenty inches in length, broad, smooth, and shiny in appearance. Its flowers appear as yellow-green spikes in early summer. It possesses a thick rhizome, which is orange beneath the rind and serves to propagate the plant. Harvested rhizomes are boiled, peeled, dried, and powdered.

The powder, imbued with a pungent smell, has a mild, warming flavor and imparts a golden color to food. The major ingredient of curry powder, the spice is employed in prepared mustard as well. The rhizomes also produce yellow or orange dyes when mordanted with chrome.

Japanese studies in the 1990s found that certain spices, including turmeric and ginger, exhibit significant antioxidant activity when added to olive, sesame, or soybean oils. In addition, turmeric has proven effective in extending the refrigerated storage life of seafoods (a fortuitous finding given that the spice has long been popular with fish preparations).

Turmeric is notable for its varied uses as a medicinal. In Samoa, the powder is mixed with lime juice and water to produce a paste that is applied to a host of skin ailments, including ulcers, the irritated navels of newborns, diaper rash, and pimples. It can also relieve the discomfort associated with dermatitis, eczema, and psoriasis as well as snakebites, insect stings, and ringworm. Combined with coconut oil, it is effective in treating more severe inflammations.

The Ayurvedic practitioners of India actively employ the herb in treating their patients. A cloth soaked in a turmeric solution can wipe away discharges of acute conjunctivitis and ophthalmia. In addition, it can dry up fluid discharges when combined with baking soda and applied to the outer ear.

The *Philippine Journal of Nursing* (50:95) states that a decoction of turmeric, followed by three glasses of water, is effective in alleviating bleeding during pregnancy (cited in Heinerman, 1996, p. 451). The addition of diced eggplant further enhances this remedy, and can be used for healing wounds.

Clinical experiments have shown that turmeric inhibits induced edema and arthritis in rodents. Its anti-inflammatory properties are comparable to those achieved by popular drugs like hydrocortisone acetate and phenylbutazone.

Rodent studies have also documented that the spice is effective in lowering serum cholesterol levels and preventing fatty accumulations in and around the liver. In 1983, the *Journal of Ethnopharmacology* (Volume 7, pages 95-109) noted that turmeric seems to be beneficial in arresting (not curing) cancer in the beginning stages of development. In 1985, *Cancer Letters* (Volume 29, pages 197-202) cited an Indian study which found that turmeric's active ingredients—curcumol and curdione—exhibited pronounced cytotoxic effects against Dalton's lymphoma cells in the early stages. The July-September issue of *Nutrition & Cancer* argued that powdered turmeric prevented the occurrence of cell mutations in association with the consumption of certain types of aggravating food (cited in Heinerman, 1996, p. 452).

A researcher at the U.S. Department of Agriculture's Nutrition Research Center in Beltsville, Maryland, Dr. Richard Anderson, has determined that extracts of turmeric, bay leaf, cloves, and cinnamon greatly enhance the production of insulin by the pancreas. Great American Natural Products, based in St. Petersburg, Florida, now markets a product called Spice Caps, which incorporates all of these ingredients. Customers have indicated that their blood sugar levels all stabilized nicely; furthermore, users who are diabetics now use less insulin than before.

Research in India has revealed that the yellow-green pigment in the spice known as curcumin stops intestinal gas in rats. As a result, experts believe that drinking turmeric powder mixed in warm water can help relieve heartburn and indigestion.

BIBLIOGRAPHY

Farrell, Kenneth T. "Turmeric." In: *Spices, Condiments, and Seasonings*. Westport, CT: AVI, 1985, pp. 221-224.

Garland, Sarah. *The Complete Book of Herbs & Spices*. Pleasantville, NY: Reader's Digest Association, 1993.

Heinerman, John. "Turmeric (*Curcuma longa* or *C. domestica*)." In: *Heinerman's Encyclopedia of Healing Herbs & Spices*. Paramus, NJ: Prentice-Hall, 1996, pp. 450-453.

"Turmeric vs. Colon Cancer." *The Herb Quarterly* (Winter 1999): 9.

V

VALERIAN

A perennial reaching two to four feet in height, valerian *(Valeriana officinalis)* thrives in uncultivated areas of Europe and the northeastern United States as far south as New Jersey and west to Ohio. Its yellow-brown, tuberous root is crowned by a hollow, angular, furrowed stem accented by dissected leaves; the latter each bear seven to ten pairs of lance-shaped leaflets.

The root possesses a number of constituents that not only produce a distinctive smell (reminiscent of dirty socks or underwear), but endow valerian with a powerful sedative character. These include valepotriates, butyl isovalerate, and eremophilene. The root extract can be taken as a tea, tablet, or capsule.

John Heinerman has designated valerian "nature's tranquilizer" (p. 458). According to C. Hobbs, in a 1996 issue of *Herbal Medicine,* it was one of the six most prescribed medicines in Europe and North America from 1733 to 1936. The root has been used as a sedative and sleep aid for several thousand years, and is widely supported by modern research as a mild sedative for the central nervous system. Russian medical publications have long touted its effectiveness in the treatment of backaches, migraine headaches, high blood pressure, and hysteria. *Planta Medica* (Volume 60, 1994, pages 278-279) posits that valerian root increases gamma-aminobutyric acid (GABA) levels in blood serum (cited in Heinerman, 1996, p. 459). GABA, considered a major inhibitory neurotransmitter within the central nervous system, is vital to glands controlled by the sympathetic nervous system, helps calm nerves, and, in high concentrations, raises brain IQ.

BIBLIOGRAPHY

Buckland, Elizabeth M. "Valerian Root." *Glenbrook Farms Herbs & Such*. www.glenbrookfarm.com. 4p.

Gardner, Jo Ann. "Valerian." In: *Living with Herbs*. Woodstock, VT: Countryman Press, 1997, pp. 261-262.

Heinerman, John. "Valerian *(Valeriana officinalis)*." In: *Heinerman's Encyclopedia of Healing Herbs & Spices*. Paramus, NJ: Prentice-Hall, 1996, pp. 458-459.

Leigh, Evelyn. "Valerian/Lemon Balm Combination Well Tolerated in Clinical Study." *HerbalGram 49* (2000): 27.

VANILLA

The universally loved vanilla flavoring is obtained, in liquid form, from unripe pods or beans growing on large green-stemmed herbaceous vines *(Vanilla planifolia)* native to the tropics, most notably Mexico, Comoro Islands, Indonesia, Malagasy Republic, Seychelles, Tahiti, Tanzania, and Uganda. A Spanish expedition led by Hernán Cortés observed the Aztecs using this flavorful spice to enhance a drink composed of ground cocoa and honey. Impressed by their flavor, the Spaniards brought the vanilla beans back to Spain where demand for them spread quickly to other European countries. Mexico continued as the sole producer of vanilla for European consumption until the mid-nineteenth century. Cultivation spread to other areas of the tropics once methods of artificial pollination and the propagation of vines by cuttings were developed.

In commercial farming endeavors, pollination is accomplished artificially (Mexico is the exception, where it is done in part artificially and in part by hummingbirds and butterflies unique to the area). Due to its high cost, vanilla extracts are now extensively adulterated. Natural vanilla accounts for roughly 5 percent of the total world market for the flavor. The comparatively inexpensive vanillan is the major flavor component in most food products (augmented by more than 150 other aroma chemicals). In the United States, where more than half the world's vanilla beans are imported, this artificial flavor commands 94 percent of the total vanilla market.

The culinary applications of vanilla—whether based on the whole vanilla bean or artificial substitutes—are far too varied to cover in one concise entry. Particularly widespread uses, however, include confections, beverages such as cocoa or coffee, cooking oil, and salad oil. Vanilla's popularity today is reflected in its inclusion in McCormick's *Flavor Trends 2000* "Top 10 Flavors for 2000."

Vanilla extract has been found effective in calming hysteria and associated emotional traumas. According to John Heinerman, the most efficient means of accomplishing this consists of placing extract-soaked cotton balls under the tongue; in this manner, the vanilla penetrates the sublingual salivary gland, traveling through the circulatory system to the brain.

In Argentina, Mexico, Venezuela, and some other Latin American countries, an alcoholic extract of the dried vanilla pods is popular as a sexual stimulant. The tincture, usually consisting of a tequila or cognac base, is generally taken in doses of ten to twenty drops, two or three times daily.

BIBLIOGRAPHY

Correll, D.S. "Vanilla—Its Botany, History, Cultivation and Economic Import." *Economic Botany* 7 (1953): 291ff.

Heinerman, John. "Vanilla *(Vanilla planifolia)*." In: *Heinerman's Encyclopedia of Healing Herbs & Spices*. Paramus, NJ: Prentice-Hall, 1996, pp. 459-461.

LaBell, F. "Vanilla—Our Most Popular Flavor." *Food Processing 49* (May 1988): 110ff.

Ranadive, Arvind S. "Vanilla—Cultivation, Curing, Chemistry, Technology and Commercial Products." In: Charalambous, George (Ed.), *Spices, Herbs and Edible Fungi.* Amsterdam: Elsevier, 1994, pp. 517-577.

W

WATERCRESS

Watercress *(Nasturtium officinale)* is a perennial plant typically found in streams and ditches filled with clear, cold water. Its creeping root system gives rise to a hollow, branching stem generally one to two feet in length. The fleshy, green leaves—odd-pinnate with one to four pairs of small, oblong (or roundish) leaflets—extend above the waterline.

Watercress is grown for its leaves, which are widely employed as garnishes and in salads. The herb's hot, tangy, and refreshing taste is enjoyed blanched and sieved as a puree or chopped and salted with brown bread and butter.

The leaves are also the basis of another well-known dish, watercress soup. This soup represents a popular age-old treatment among the Chinese for canker sores on the tongue or lips, bad breath, mouth blisters, and swollen gums.

Watercress has many other medicinal applications as well. A tea or juice extract is effective in eliminating accumulated fluids in body tissue (e.g., gout) and for ridding the lungs of mucous congestion. A solution—made from the leaves combined with boiled apple cider, which is left to stand, and then strained—is applied in a cloth to the forehead to cure headaches resulting from illness or stress. An infusion of watercress (obtained by boiling the leaves in water and refrigerating the strained liquid) has long been popular for curing eczema and dermatitis.

Watercress contains a high proportion of Vitamin C and vital minerals such as iron. Therefore, it is a natural tonic with a proven track record as a remedy for anemia, rickets, weak heart and eyesight, and diminished milk flow.

BIBLIOGRAPHY

Garland, Sarah. *The Complete Book of Herbs & Spices*. Pleasantville, NY: Reader's Digest Association, 1993.

Heinerman, John. "Watercress *(Nasturtium officinale)*." In: *Heinerman's Encyclopedia of Healing Herbs & Spices*. Paramus, NJ: Prentice-Hall, 1996, pp. 464-466.

WILLOW BARK

Willow bark has been employed for a variety of medicinal purposes since antiquity. In ancient Greece, arthritic joints were immersed in clay pots containing willow bark or leaves steeped in olive oil. Native Americans across North America used the bark or leaves to relieve pain, inflammation, and fever. Ingestion took many forms, including wearing poultices, drinking root tea, and chewing leaves as well as rubbing them directly on the body. American folk medicine also utilized the bark as a blood thinner and to treat fevers.

While the willow family includes a considerable number of species (the *Salix* genera alone encompasses approximately 300), only the following are widely recognized as sources of medicinal bark: the European white willow (*Salix alba;* used since the beginnings of recorded history), crack willow *(Salix fragilis)*, purple willow *(Salix purpurea)*, violet willow *(Salix daphnoides)*, bay willow *(Salix pentandra)*, and the black willow (*Salix nigra;* most widely used by Native Americans).

The inner bark of both willows and poplars (genera *Populus*) contains phenolic glycoside esters (generally in higher concentrations during the spring and summer months). These compounds are transformed into saligenin by intestinal microorganisms. The saligenin is then oxidized in the liver and blood, producing salicylic acid. Salicylic acid reduces pain by inhibiting the synthesis of prostaglandins in sensory nerves.

The United States permits the production of 270-mg capsules of willow-bark extract standardized to 41 mg of salicin. The powdered variant sold in health food stores for brewing teas is considered too weak to be medicinally effective; for example, between three pints and five quarts of this concoction would have to be ingested daily in order to equal the 4.5 grams of aspirin typically taken to counter arthritis pain. Chief risks include upset stomach and, in situations where children have chicken pox or the flu, Reye's syndrome.

BIBLIOGRAPHY

Foster, Steven. *Herbs for Your Health.* Loveland, CO: Interweave Press, 1996.

Foster, Steven. "Plants for Pain Relief." *The Herb Companion* (August/September 1997): 60-61.

Leake, C.D. *An Historical Account of Pharmacology to the Twentieth Century.* Springfield, IL: Charles C Thomas, 1975.

Tyler, V.E. *Herbs of Choice: The Therapeutic Use of Phytomedicinals.* Binghamton, NY: Pharmaceutical Products Press, 1994.

Tyler, V.E. *The Honest Herbal,* Third Edition. Binghamton, NY: Pharmaceutical Products Press, 1993.

WITCH HAZEL

Witch hazel (genus *Hamamelis*) is one of approximately 100 plant groups found only in both eastern North America and eastern Asia. Two species of witch hazel are native to North America while two to four species are found in Asia. The witch hazel long sold in U.S. pharmacies as an astringent is derived from steam-distilling the twigs of *Hamamelis virginiana* and adding alcohol as a preservative. In Europe, the extract contains both the twigs and leaves of the plant; it is employed not only as an astringent but as an anti-inflammatory to treat minor skin injuries, eczema, burns, varicose veins, and other topical ailments.

Although little scientific evidence exists regarding its value in such products, researchers based in Yokkaichi, Japan, determined in a 1996 study that witch hazel exhibited strong antioxidant activity in skin tissue. In other words, by protecting skin cells against destructive oxygen molecules called radicals, the plant would appear to aid in the prevention of aging or wrinkling of the skin.

Witch hazel's healing powers were once attributed to tannins, compounds widely found in other plants. However, researchers at the Dr. Willmar Schwabe Group and the Johann Wolfgang von Goethe University in Germany determined that a fraction of the European extract composed primarily of proanthocyanidins inhibited topical viruses such as herpes simplex as well as reducing inflammation to a greater degree than did fractions high in specific tannins.

BIBLIOGRAPHY

Garland, Sarah. *The Complete Book of Herbs & Spices*. Pleasantville, NY: Reader's Digest Association, 1993.

"Witch Hazel's Healing Powers." *The Herb Companion* (December/January 1996/1997): 70.

"Wrinkle Work." *The Herb Companion* (August/September 1996): 72.

WOOD BETONY

A perennial herb reaching as tall as three feet in height, wood betony *(Stachys officinalis)* is typically found in meadows, hilly slopes, and forests. Its stem is hairy, unbranched or slightly branched, and quadrangular; the leaves—minty in aroma—are also hairy and positioned opposite one another, with those on top lanceolate and the bottom ones oblong-cordate in shape. Red-purple flowers bloom during the summer months.

It is a member of the betony family, commonly known as woundworts due to their long tradition of healing. Notable examples include lamb's ears (woolly betony) and the great-flowered betony.

Though today wood betony is used primarily to flavor herb teas, in the past it was a very popular remedy with herbalists. The flowers, leaves, and stems of the plant were brewed as a treatment for bladder problems, edema, gout, intestinal worms, kidney dysfunction, and stomach problems. Both tea and herbal formulas in capsule form are still used today for combating allergies and headaches as well as normalizing blood pressure.

BIBLIOGRAPHY

Gardner, Jo Ann. *Living with Herbs.* Woodstock, VT: Countryman Press, 1997.

Heinerman, John. "Wood Betony *(Stachys officinalis).*" In: *Heinerman's Encyclopedia of Healing Herbs & Spices.* Paramus, NJ: Prentice-Hall, 1996, pp. 476-477.

WORMWOOD

A perennial herb reaching one to three feet in height, wormwood *(Artemisia absinthium)* features grayish-white stems covered with fine silky hairs that grow in a shrublike pattern. The yellow-green leaves are silky, hairy, and glandular, and are accented with small resinous particles. The leaves and flowering tops (both in fresh and dried form) constitute the usable portion of the plant. It has an aromatic odor and spicy, rather bitter, taste.

Originally found in Europe, northern Africa, and western Asia, wormwood is now widely cultivated throughout the world. Wormwood plays a role in the production of vermouth. A mixture of its crushed leaves and apple cider vinegar also makes an excellent insect repellent.

A blend of wormwood and whiskey has long been used as a pain reliever, most notably for labor during pregnancy, tumors and cancers, aching muscles, arthritic joints, sprains, dislocated shoulders and knees, and fractured bones. Wormwood oil can also be applied externally in much the same manner. Joseph Smith Jr., eldest son of the famed Mormon leader, left a written account of the treatment when a carriage door crushed two of his fingers:

> Taking from her trunk a little bottle of whiskey and wormwood, [mother] turned the tips of my fingers upward, and poured the liquid upon them, into the dressings—at which, for the first time

> in my life I promptly fainted! It seemed as if she had poured the strong medicine directly upon my heart, so sharply it stung and so quick was its circulatory effect. . . . I soon recovered and we proceeded on our journey, reaching home in good time and without further mishap. (Heinerman, 1996, p. 478)

A study outlined in *Planta Medica* (Volume 37, pages 81-85) indicated that various wormwood species have been utilized clinically to treat hepatitis as well as protect the liver from lesions resulting from the ingestion of harmful chemicals. The *Chemical and Pharmaceutical Bulletin* (Volume 31, page 352) touted wormwood as an important remedy for jaundice and inflammation of the gallbladder (cholecystitis). In these applications the herb can be ingested as either a tea or powdered capsule (Heinerman, 1996, p. 479).

A related species, sweet wormwood, is popular as an ornamental; its essential oil can be used in antibacterial and antifungal applications. The *Zhou Hou Bei Ji Fang (A Handbook of Prescriptions for Emergencies),* by Chinese herbalist Ge Hong (A.D. 231-341) states, "Take a handful of sweet wormwood, soak it in a Sheng [roughly one liter] of water, squeeze out the juice and drink it all for malaria." Recent tests at the Institute of Chinese Materia Medica of the China Academy of Traditional Chinese Medicine determined the herb to be effective in killing malaria parasites.

BIBLIOGRAPHY

Gardner, Jo Ann. "Wormwoods." In: *Living with Herbs*. Woodstock, VT: Countryman Press, 1997, pp. 265-268.

Heinerman, John. "Wormwood *(Artemisia absinthium).*" In: *Heinerman's Encyclopedia of Healing Herbs & Spices*. Paramus, NJ: Prentice-Hall, 1996, pp. 477-479.

Y

YARROW

Yarrow *(Achillea millefolium)* has been used throughout civilization in most corners of the world. Consequently, it has come to be known by a variety of names, including milfoil, nosebleed, soldier's woundwort, and staunchweed. The first word of the botanical name is derived from its use in the Trojan War by Greek warrior Achilles to stop bleeding from battle wounds; the second word denotes the finely feathered character of the leaves.

A hardy perennial, yarrow thrives from seed or root in most climates; in many regions, it is considered a weed. The plant's downy stems can reach two feet in height. Small, gray-white (sometimes pink) flowers appear in clusters throughout the summer months. (The cultivated variety grown as a garden ornamental is deep yellow in color.)

The uses of yarrow are extremely diversified, spanning gardening, cooking, beer brewing, medicine, and homeopathy. Its young leaves are tasty chopped in salads or boiled like spinach. It is substituted for hops (resulting in a much stronger brew) in Sweden to make beer. In gardens, it allegedly increases the health of surrounding plants. It encourages fermentation in composting, and is valued as a fertilizer for soils (particularly those deficient in copper).

Yarrow has a long history in invoking black magic and as a protection against evil. Sarah Garland notes that "it was used as a herb of divination by the early Chinese—whose ancient book, *I Ching,* is also known as *The Yarrow Stalk Oracle*—and by the Druids and Anglo-Saxons" (p. 24).

The entire aboveground portion of the herb has medicinal value. Its active ingredients include lactones, flavonoids, tannins, coumarins, saponins, sterols, achilleine, cyanidin, amino acids, salicylic acid, sugars, and volatile oil composed of azulene, borneol, terpineol, camphor, cineole, isoartemesia ketone, and thujone.

The dark blue essential oil—derived from a steam distillation of the flowers—possesses an anti-inflammatory action. It is combined with peppermint, eucalyptus, or thyme oils to treat colds and flu with chest congestion. When mixed with St.-John's-wort oil, it can relieve rheumatic pains.

Yarrow flower essence is used to terminate internal bleeding or heavy menstrual periods. Astringent in nature, the crushed yarrow leaves constrict blood vessels locally, resulting in its use as a styptic for cuts and nosebleeds. A tea extract can bring down cold and flu fevers, lower blood pressure, reduce blood clots, treat dysmenorrhea and amenorrhea, and combat mucus build-up caused by either poor digestion or allergies.

A number of problems are associated with the plant. Its use can cause rashes and increased sensitivity to the sun. It can have an additive effect with other blood pressure–lowering herbs or medications, including diuretics. Taken in large doses, yarrow is a uterine stimulant; therefore, it should be avoided while pregnant.

BIBLIOGRAPHY

Gardner, Jo Ann. "Yarrow (Milfoil)." In: *Living with Herbs*. Woodstock, VT: Countryman Press, 1997, pp. 269-271.

Garland, Sarah. *The Complete Book of Herbs & Spices*. Pleasantville, NY: Reader's Digest Association, 1993.

Langer, Robin E. "Herb-of-the-Month: Yarrow *(Achillea millefolium)*—July 2000." *Snowbound Herbals* (September 6, 2000): 2p. www.sbherbals.com.

YELLOW DOCK

A perennial reaching one to three feet in height, yellow dock *(Rumex crispus)* is widely held to be an undesirable weed proliferating in fields and waste areas across Europe and North America. It possesses a spindle-shaped, yellow taproot that nurtures a smooth, slender stem punctuated by oblong-lanceolate–shaped, pointed leaves. Light green in color with wavy margins, the leaves are larger and longer-petioled in the lower portion of the plant. The numerous, pale-green, drooping flowers produce a pointed, three-angled, heart-shaped nut or seed.

A decoction made from the root—to be ingested or applied directly to the skin—serves as a blood purifier for most eruptive diseases. Yellow dock tea also improves digestion and stimulates the liver and colon.

Yellow dock syrup—produced by boiling the root in water and adding the stained liquid to dark honey, blackstrap molasses, and small amounts of maple syrup and vanilla—is an effective remedy for upper respiratory problems (e.g., emphysema). Because they are richer in vitamin A than carrots, yellow dock greens are considered useful for improving night vision, particularly for eyes that have become dimmed with age.

BIBLIOGRAPHY

Heinerman, John. "Yellow Dock *(Rumex crispus).*" In: *Heinerman's Encyclopedia of Healing Herbs & Spices*. Paramus, NJ: Prentice-Hall, 1996, pp. 482-483.

Appendix

Supplementary List of Botanicals and Their Applications

Aconite	Arthritis and gout; homeopathy
Adder's-tongue	Wounds; sciatica and neuralgia; respiratory system problems
Agave	Spasms; coughs; poor urination; accumulated phlegm; fevers; kidney inflammation; skin sores; urinary tract infection; wounds
Agrimony	Emotional hypersensitivity; itchy eyes; hives and moist skin eruptions; asthma and bronchitis
Allspice	Toothache; bad breath
Amaranth	Nutrition; diarrhea; bleeding
American Centaury	Freckle removal
Angelica	Menopause; PMS
Annatto	Intestinal gas and heartburn; indigestion
Apple Tree Bark	Fevers; gallbladder attacks; delayed menstruation; nausea and vomiting; liver and spleen disturbances; abdominal cramps; kidney stones; Bright's disease; dysentery; boils; insect bites; rabies; toothaches
Arrowroot	Diarrhea; erysipelas; sunburns
Artichoke	High cholesterol; liver problems
Arugula	Flavoring
Arum	Sore throat; ringworm
Asafoetida	Stomach pains; headaches; snakebites; flavoring
Ash	Mastitis; splenitis; atopic dermatitis; gout and arthritis; intestinal parasites
Asparagus	Blemishes; kidney problems; nutrition
Aster	Hemorrhaging; excessive menstruation; bloody stool; diarrhea
Astragalus	Physical exhaustion; disease prevention; cancer tumors
Atractylis	Diabetes; hemorrhages

Azalea	Lice; ringworm; chiggers
Bamboo	Radiation exposure; impotence
Barberry	Fevers; constipation; alcoholic hangovers; hepatitis
Barley Grass	Gluten allergies; low energy level
Bay	Dandruff; bronchitis; coughs; arthritis
Bayberry Bark	Varicose veins; fevers; parasites; constipation; prolapsed uterus; excessive menstruation; vaginal discharge
Bearded Darnel	Herpes; bedsores; dizziness; insomnia
Beech	Frostbite; scalds; burns; poison ivy/oak; diaper rash; gangrene
Beetroot	Cancer
Bergamot	Nausea; fevers and colds; migraines
Birch	Wounds; sores; diarrhea; cancer
Bird-of-Paradise	Insect stings
Birthwort	Bacterial infection
Bitters	Cardiac irregularities; gastritis; gallbladder problems
Black Alder	Sore throats
Blackberry	Diarrhea; blood impurities; sluggish liver; mouth and skin sores; bad breath
Black Haw	Miscarriage; asthma
Bladder Wrack	Goiter; glandular insufficiency
Blazing Star	Venereal disease; sore throats
Blessed Thistle	Insufficient breast milk; liver problems; digestive disturbances
Bloodroot	Cancer
Blue Cohosh	Menstruation; labor problems
Blue Flag	Staph infections
Blue Spruce	Mouth sores; scurvy
Blue Vervain	Hyperactivity; colds and flu
Bluebottle	Black eye and bruises
Borage	Depression; fevers; swellings
Box Elder	Mild burns and sunburn
Bracken Fern	Lung problems
Brier Rose	Illness recuperation; wounds; constipation
Broom	Edema
Bryony	Whooping cough; cardiac disorders
Buchu	Gout; kidney stones; urine retention; prostate problems; bladder infection; bruises
Bugleweed	Nosebleed; excess menstruation; hemorrhoids; bloody urine and stool; labor recuperation

Bupleurum	Prolapsed uterus; headaches; backaches
Buttercup	Shingles; sciatica
Butternut	Dysentery
Cactus	Diabetes; dry, itchy scalp; wounds
Camphor	Infectious diseases; toothaches; backaches; sore muscles; sciatica; rheumatism; opium withdrawal; asthma; bronchitis; respiratory disorders; claustrophobia
Caraway	Delayed menstruation; heartburn and acid indigestion; flavoring
Cardamom	Celiac disease; indigestion; flavoring
Cascara Sagrada	Constipation; herpes simplex
Castor Bean	Respiratory problems; skin diseases
Catnip	Insomnia; hyperactivity; aching teeth and gums; eye inflammation; dizziness
Cattail	Diarrhea; poison ivy rash; burns; insect bites; toothaches; cuts; hair problems; stress; fatigue; poor appetite
Celandine	Foot calluses
Celery	Nervousness; insect stings; weight problems; nutrition
Chicory	Fertility; jaundice and other liver/spleen problems; high cholesterol; rapid heartbeat; acid indigestion; gallstones; flavoring
Chinese Cucumber	Dehydration; glandular inflammation; poor urination; asthma; bronchitis; measles; mumps; chicken pox; laryngitis; abscesses; breast inflammations; AIDS
Chives	Anorexia nervosa
Chrysanthemum	Hypertension; angina pain; "feverish air"
Cilantro	Food poisoning; flavoring
Cinchona	Fevers; flu; neuralgia; headache; sore throats; poor appetite
Cinquefoil	Sore throats; tonsillitis; inflamed gums
Citrin	Obesity
Clary Sage	Frigidity and impotency in later years; sore throats; laryngitis
Cleavers	Loose skin; epileptic seizures
Club Moss	Skin problems; fatigue; homeopathy
Coca	Physical pain; fatigue; flavoring
Coltsfoot	Respiratory congestion; inflammations; asthma
Corn silk	Kidney problems; cystitis; pyelitis; oliguria; edema

Corydalis	Parkinson's disease; palsy; general trembling; depression
Costmary	Liver and stomach disorders; childbirth complications; odors; flavoring
Cotton	Delayed menstruation; birth control; bleeding during childbirth; herpes
Cumin	Abdominal pain; flavoring
Daffodil	Swellings and dislocations; muscle pulls and sprains; bruises
Dandelion	Warts; liver problems; hypertension; diabetes; night blindness; childhood fevers
Dogwood	Malaria
Dong Quai	Gynecological problems; physical pain; cardiovascular problems
Elderberry	Mucus accumulation; constipation; migraines; stomachaches; eruptive sores and burns; flavoring
Elecampane	Rheumatism; sciatica; lumbago; facial neuralgia; asthma; bronchitis; chest colds; tuberculosis; pneumonia
Elephant Grass	Erysipelas; seizures; wounds; sores; gangrene
Eyebright	Eye and throat ailments
Fungus	Cancer; chronic fatigue syndrome
Gentian	Low stamina
Gladiola	Dysentery; diarrhea; colds
Gotu Kola	Mental retardation; phlebitis; varicose veins
Guar Gum	Obesity; high cholesterol; diabetes
Guarana	Low stamina
Hawthorn	Hypertension; coronary problems
Horehound	Mucus congestion
Horseradish	Muscular aches; skin conditions; cold sensations; postnasal drip; flavoring
Ho-Shou-Wu	Impotency
Ipecacuanha	Poisoning
Iris	Coughs and hoarseness
Japanese Honeysuckle	Nervous tension and anxiety
Jasmine	Sexual frigidity
Jewelweed	Poison ivy/oak
Kola Nut	Fatigue; drug withdrawal; birth control
Lady's Mantle	Eye inflammations; rough skin; decoration
Lilac	Acid indigestion; fevers; gout; abdominal cramps; rheumatism

Lily	Upset stomach; trauma; poison ivy; wounds; bruises
Lovage	Abdominal pains; flatulence
Magnolia	Heart disease; hypertension; cancer
Maple	Boils; sores; wounds; flavoring
Marjoram; Oregano	Fevers; cramps; epilepsy; irregular menstruation; childhood diseases
Marshmallow	Hernias; cystitis
Mistletoe	Hypertension; high blood pressure
Morning Glory	Headache; inflamed eyes; insect bites; bathing sores; boils
Moss	Excessive bleeding
Mountain Mahogany	Prostatism; constipation; hemorrhoids; common cold; influenza; conjunctivitis
Mullein	Asthma; coronary distress; intestinal infection; tonsillitis; mumps; chicken pox; measles; earache; skin problems
Mustard	Knee and joint pain; lower backache; chest colds; deep coughs; flavoring
Myrrh	Bad breath; gingivitis; gum inflammation
Nasturtium	Mucus congestion; flavoring
Neem	Scabies; fungus; mosquito bites
Nigella	Decoration; flavoring
Nux Vomica	Opium and heroin addiction; drug overdose; hangovers; cardiac arrest
Oak	Medication poisoning; varicose veins; bloody stool/urine; fevers; hemorrhoids; excessive menstruation; mouth sores; skin irritations; sores; sore throat
Olive Leaf	Cancer
Osha	Viral infections; allergies; asthma; bronchitis; indigestion
Passionflower	Hyperactivity; drug addictions
Pennyroyal	Scabies; psoriasis; hives; shingles; measles; mumps; chicken pox; diaper rash; poison ivy and oak
Peony	Gout; asthma
Pine	Acne; blood clotting; bronchitis; high cholesterol; diabetes; emphysema; influenza; leukemia; lupus; migraines; morning and motion sicknesses; paralysis; Parkinson's disease; rheumatoid arthritis; scurvy; skin ulcers; sinusitis; stroke; varicose veins

Plantain	Toothaches; peptic ulcers; lacerations
Pokeroot	Chronic skin diseases; glandular swelling
Poplar	Cancer; skin ulcers; gangrenous wounds; eczema; burns; strong perspiration
Prickly Ash	Paralysis; toothache pain; inactive salivary glands; peritonitis; distention of the bowels; abdominal inflammation; cholera; typhus; typhoid; pneumonia; sickle-cell anemia
Pumpkin	Tapeworms
Purslane	Respiratory disorders; skin afflictions; hypotension; urinary tract disorders; nutrition
Pyrethrum	Fly control
Ramps	Aging; earaches; croup; colds; sore throats; cold sores; toothaches; intestinal parasites
Red Raspberry	Flu; common cold; childbirth complications
Rhododendron	Chronic cystitis; mucus accumulations; kidney problems
Roses	Infection; skin wrinkling; minor cuts; bruises; sprains; ligament pulls; sore eyes; mouth and throat inflammation; stomach ulcers; nervousness; liver problems; rheumatism
Sagebrush	Diarrhea; dehydration; sunstroke; eye inflammation; wounds; early hair loss; dandruff; ringworm; eczema; psoriasis; blood poisoning
Senna	Constipation
Shepherd's Purse	Excessive bleeding
Skullcap	Nervous spasms or convulsions; high cholesterol; arthritis
Slippery Elm	Sores; eczema; fevers; cigarette cravings
Snapdragon	Sore eyes
Spruce	Headaches; coughs; common cold; flu
Squawvine	Childbirth complications; syphilis; gonorrhea; yeast infection; insomnia
Suma	Anemia; PMS; cancer prevention
Sumac	Venereal disease; poison ivy
Sweet Cicely	Heartburn; acid indigestion; abdominal cramps; intestinal gas; poor appetite
Tarragon	Insomnia; hyperactivity; indigestion; poor appetite; gout; rheumatism; urine retention; sluggish kidneys and bladder; irregular periods
Uva Ursi	Kidney problems; childbirth complications; cardiac edema; allergies

Walnut	Diarrhea; sore throats; ringworm; runny nose; head colds; acne; eczema; psoriasis; heart disease
Wheat Grass	Flagging energy; nutrition
White Oak	Hemorrhoids; excessive menstruation; blood urine; internal hemorrhaging; body odor; diarrhea
Wild Black Cherry	Coughs; poor appetite
Wild Mexican Yam	Rheumatism; aging; excessive weight gain
Wild Oregon Grape	Spinal meningitis; insect stings; snakebites
Wintergreen	Fevers; sore throats; headaches
Yohimbine	Sexual dysfunction; obesity
Yucca	Cancer; arthritis
Zedoary	Sluggish liver; constipation; stomach problems; gallbladder attacks; cuts; sprains; varicose veins

Recommended Reading

American Herbal Association. *Complete Book of Herbs: Using Herbs to Enrich Your Garden, Home and Health.* Lincolnwood, IL: Publications International, 1997.

Anderson, Frank J. *An Illustrated History of the Herbals.* New York: Columbia University, 1977.

Arber, Agnes R. *Herbals: Their Origin and Evolutions: A Chapter in the History of Botany, 1417-1670,* Third Edition. Cambridge, England: Cambridge University, 1988. Revision of first 1912 ed.

Bardswell, Frances A. *Herb-Garden,* Second Edition. London: Bracken, 1986, c.1911.

Blunt, Wilfrid and Raphael, Sandra. *The Illustrated Herbal,* Revised Edition. London: Frances Lincoln, 1994.

Bown, Deni. *Encyclopedia of Herbs and Their Uses.* London: Dorling Kindersley, 1995.

Bremness, Lesley. *The Complete Book of Herbs and Spices.* New York: Viking Penguin, 1995.

Brownlow, Margaret E. *Herbs and the Fragrant Garden,* Third Edition. London: Darton, Longman, and Todd, 1979.

Clarkson, Rosetta E. *Green Enchantment.* New York: Collier, 1991, c.1940.

Clarkson, Rosetta E. *Herbs: Their Culture and Uses.* New York: Wings, 1996. Originally published: New York: Macmillan, 1942; earlier version privately published in 1920.

Clarkson, Rosetta E. *Magic Gardens.* New York: Collier/Macmillan, 1992, c.1939.

Clevely, Andi M. and Richmond, Katherine. *The Complete Book of Herbs.* New York: Hermes House, 1999.

Cuthbertson, Yvonne. *Women Gardeners: A History.* Denver, CO: Arden, 1998.

Foster, Steven. *Herbal Renaissance: Growing, Using and Understanding Herbs in the Modern World.* Salt Lake City, UT: Gibbs-Smith, 1993.

Freke, Ellen. *The Complete Book of Spices.* London: Apple, 1997.

Garland, Sarah. *The Complete Book of Herbs and Spices.* Pleasantville, NY: Reader's Digest Association, 1993.

Gordon, Lesley. *A Country Herbal.* New York: Mayflower, 1980.

Grieve, Maud and Leyel, Hilda (Eds.). *A Modern Herbal: The Medicinal, Culinary, Cosmetic, and Economic Properties, Cultivation, and Folklore of Herbs, Grasses, Fungi, Shrubs, and Trees with All Their Modern Scientific Uses.* 2v. New York: Hafner, 1967, c1931.

Griggs, Barbara. *Green Witch: A Modern Woman's Herbal.* London: Vermilion, 2000.

Hawkins, Kathryn and Duff, Gail. *A Dash of Spices: Spices for Beauty, Health and Cooking.* Edison, NJ: Chartwell Books, 1997; London: Reader's Digest, 1997.

Hobbs, Christopher. *Herbal Remedies for Dummies.* Foster City, CA: IDG; London: Transworld, 1999.

Hollis, Sarah. *Country Diary Herbal.* New York: Henry Holt, 1990.

Hopkinson, Simon G. and Hopkinson, Judith. *Herbs.* Chester, CT: Globe Pequot, 1989.

Kowalchik, Claire and Hylton, William H. (Eds.). *Rodale's Illustrated Encyclopedia of Herbs.* Emmaus, PA: Rodale, 1987.

Lucas, Richard. *Common and Uncommon Uses of Herbs for Healthful Living.* New York: Arco Publishing, 1969.

McCaleb, Robert S., Leigh, Evelyn, and Morien, Krista. *Encyclopedia of Popular Herbs: Your Complete Guide to the Leading Medicinal Plants.* Roseville, CA: Prima, 2000.

Mulherin, Jennifer. *Macmillan Treasury of Spices and Natural Flavorings.* New York: Macmillan, 1988.

Robbins, Christopher. *Thorsons' Introductory Guide to Herbalism.* San Francisco, CA: Thorsons, 1993.

Rohde, Eleanour S. *A Garden of Herbs,* Revised Edition. London: Jenkins, 1933.

Rohde, Eleanour S. *Herbs and Herb Gardening.* London: Medici Society, 1946, c1936.

Rohde, Eleanour S. *Old English Herbals,* Third Edition. London: Minerva, 1974.

Sanecki, Kay N. *The Complete Book of Herbs.* New York: Macmillan, 1974.

Sanecki, Kay N. *History of the English Herb Garden.* London: Ward Lock, 1992.

Shaudys, Phyliss. *Herbal Treasures: Inspiring Month-By-Month Projects for Gardening, Cooking, and Crafts.* Pownal, VT: Storey Communications, 1990.

Smith, Keith V. *Illustrated Earth Garden Herbal: A Herbal Companion,* Revised Edition. Port Melbourne, Victoria, Australia: Thomas C. Lothian, 1994.

Stuart, Malcolm (Ed.). *Encyclopedia of Herbs and Herbalism.* New York: Grosset and Dunlap, 1979.

Swahn, Jan O. *The Lore of Spices: Their History and Uses Around the World.* New York: Barnes and Noble Books, 1997, c1991.

Thomson, William A.R. (Ed.). *Healing Plants: A Modern Herbal.* London: Macmillan, 1980.

Thomson, William A. R. and Smith, Elizabeth. *Healing Herbs.* London: BBC, 1978.

Tolley, Emelie and Mead, Chris. *Herbs: Gardens, Decorations, and Recipes.* New York: Clarkson Potter, 1985.

Touissant-Samat, Magaglonne. *History of Food.* Oxford, England: Blackwell, 1992.

Whiteman, Robin. *Brother Cadfael's Herb Garden; An Illustrated Companion to "Medieval Plants and Their Uses."* Photography by Robin Talbot. Boston: Little, Brown, 1997.

The following works by John Heinerman merit special mention for their valuable contributions to the field of herbal medicine:

Dr. Heinerman's Encyclopedia of Anti-Aging Remedies. Paramus, NJ: Prentice-Hall, 1997.

Dr. Heinerman's Healthy Homestyle Cooking. Paramus, NJ: Prentice-Hall, 1998.

Family Encyclopedia of Natural Healing. Springville, UT: Cedar Fort, 2000.

Folk Medicine in America Today. New York: Kensington, 2001.

The Healing Benefits of Garlic: From Pharaohs to Pharmacists. New Canaan, CT: Keats, 1994.

Healing Power of Herbs. Boca Raton, FL: Globe Communications, 1997.

The Health Benefits of Cayenne. New Canaan, CT: Keats, 1997.

Heinerman's Encyclopedia of Healing Herbs & Spices. West Nyack, NY: Parker, 1995.

Heinerman's Encyclopedia of Healing Juices. West Nyack, NY: Parker, 1994.

Heinerman's Encyclopedia of Juices, Teas and Tonics. Englewood Cliffs, NJ: Prentice-Hall, 1996.

Heinerman's Encyclopedia of Nature's Vitamins and Minerals. Paramus, NJ: Prentice-Hall, 1998.

Heinerman's New Encyclopedia of Fruits and Vegetables (Revised and expanded). West Nyack, NY: Parker, 1995.

Miracle Healing Foods (Revised edition). Paramus, NJ: Prentice-Hall, 1999.

Index

Page numbers followed by the letter "b" indicate boxed text.